SYSTEM SAFETY: TECHNOLOGY AND APPLICATION

GARLAND SAFETY MANAGEMENT SERIES

Dan Petersen — Series Advisor

ANALYZING SAFETY PERFORMANCE   *Dan Petersen*

HUMAN-ERROR REDUCTION AND
SAFETY MANAGEMENT   *Dan Petersen*

IMPROVING PERFORMANCE FOR
SAFETY AND HEALTH   *Kingsley Hendrick*

THE MEASUREMENT OF SAFETY
PERFORMANCE   *William E. Tarrants*

QUANTITATIVE INDUSTRIAL   *Harry J. Beaulieu*
HYGIENE   *Roy M. Buchan*

SOURCEBOOK ON ASBESTOS   *Barbara Peters*
DISEASES   *George Peters*

Forthcoming Titles

Computerizing Safety Data   *Charles Ross*

Respiratory Protection Handbook   *John Pritchard*
                                       Allen Lilly

Safety Communications   *Lawrence E. Schlesinger*

SYSTEM SAFETY: TECHNOLOGY AND APPLICATION

Second Edition

Sol W. Malasky

Adjunct Faculty
System Safety Department
University of Southern California

Garland STPM Press
New York & London

Library of Congress Cataloging in Publication Data

Malasky, Sol W.
    System safety.

    (Garland safety management series)
    Includes index.
    1. Product safety.   2. System design.   I. Title.
II. Series.
TS175.M34   1981   658.5'6   81-6475
ISBN 0-8240-7280-4   AACR2

Published by Garland STPM Press
136 Madison Avenue, New York, New York 10016

Printed in the United States of America

15   14   13   12   11   10   9   8   7   6   5   4   3   2   1

To My Wife and Children

Contents

Preface to the Second Edition

The first edition of this book commented on the increasing rate at which knowledge and technology are growing. Indeed, considerable change and growth has taken place in system safety during the years that has passed since that edition's publication. At that time, and even until just a few years ago, the attitude generally held was that government help was sorely needed to diminish the number of accidents and fatalities occurring each year. It was noted for example, that health care improvement had progressed to the point that accidents had become the most prominent cause of death from ages 10 to 45. Further, an increasing number of deaths above age 45 were being attributed to low-level, long-term exposure to toxic agents and carcinogens. Today, claims are made by some that efforts spearheaded by government to assure safety at home, in the workplace, and on the highway, for example, have been carried too far and that some retrenchment is in order. In particular, it is suggested that some of the work expended to achieve safety in such areas places a drag on productivity and, consequently, helps fuel inflation, one of the nation's greatest concerns. Those in favor of continued or increased government intervention in safety matters suggest that there is no drag on productivity, but rather the gross national product (GNP) is deficient as a measure of work expended. One solution suggested is to make the GNP a part of a more comprehensive measure, one that reflects quality of life, such as Japan's Net National Welfare or the Overseas Development Council's Physical Quality of Life Index (PQLI). In this way, it is argued, certain labors now not counted in the GNP, particularly some associated with enhancing safety, would have their usefulness measured.

Another significant change that has occurred since the first edition is the increase of litigation as a remedy for injuries suffered as a result of defective design, manufacture, or labeling. Recent court decisions have expanded the number of people in the chain from designer to retailer who are liable for such injuries, and recent jury verdicts have greatly increased the size of the awards made to plaintiffs in such cases. Here, too, there are some who consider that the pendulum has swung too far, and various laws are being debated at national and state levels to place some limits upon such liability. These include limits on the amount of money that may be sought by plaintiffs and upon the length of time after purchase that an aggrieved user may seek redress.

Advances have been made across a broad spectrum of technological areas of system safety since the first edition. Improvements made in the ability to predict hazards and their likelihood of occurrence are being used in the development of new products, from children's toys to esoteric scientific equipment, and in new processes used in factories and for leisure-time activities. System safety now makes greater use of computers for carrying out studies not otherwise possible and for conducting routine tasks that would otherwise require much time.

Treatment of these aspects of system safety previously mentioned—ethical considerations, product liability, and recent technological advances—has been expanded in this revision. Certain material discussed in the first edition remains inherently valid and required less updating, such as the optimization process needed to balance the various requirements levied on a system with those specifically established to assure safety, the management and systems methodologies useful for accomplishing such optimization, the interfaces between system safety and allied disciplines, and the underlying Boolean algebra and statistics useful in conducting systems safety studies.

Preface to the First Edition

The notion of safety, always of concern to man, has in recent times taken a position of greater importance in our everyday lives. Two aspects are uppermost. The first relates to considerations of ethics. What is the responsibility of the developer or manufacturer of a new product with regard to safety? To what extent is the physician or the pharmaceutical manufacturer liable for a statistically rare side effect exhibited by a drug that has been approved for use by the Federal Food and Drug Administration?

The second aspect of safety is technical in nature. Much of the recent evolution in the technical aspects of system safety has resulted from efforts undertaken by the aerospace industry. In particular, considerable system safety technology was developed to assure the safety of missiles carrying nuclear warheads and of personnel engaged in manned space exploration. In the here and now, the growth of consumerism has brought about an almost universal concern with safety aspects of almost all equipment and activities that touch upon human affairs—automobiles, food, clothing, toys, cosmetics, and our ecological envelope. One consequence of this new-found concern with safety on the part of so many is evidenced by an increase in the number of regulatory bodies at all governmental levels and the number of regulations promulgated by these organizations. Perhaps the most difficult aspect of attempts to regulate safety in areas of concern to the average citizen is in the interrelationship of the two components of safety, ethical and technical.

As a further consequence of this new wave of consumerism, many in industry and government outside the aerospace industry, now find it necessary to relate more extensively and more formally to system safety considerations. Those in industry who have been conducting safety programs as a natural part of their activities now find it necessary to supply safety documentation in response to the increasing regulation in this field. It is hoped that this book will be useful in providing those individuals whose work or study requires that they interface with hardware or human systems with a working knowledge of the technical aspects of system safety and an understanding of how to develop and carry out a system safety program.

Since no comprehensive, organized presentation could be found on the subject of system safety at the time this book was initiated, it was considered desirable that the main exposition be extensive and complete. The first few chapters touch upon the ethics of system safety, discuss the language of system safety, and describe how system safety relates to other disciplines. The next few chapters deal with the kernal of system safety which is involved with the identification and control of those hazards that pose a threat to man or the systems with which he interfaces. In some of these interfaces, particularly those in which the state of the art is being rapidly advanced, the problem is made more complex because all aspects and implications of the hazards involved are not yet fully understood.

First steps generally taken in hazard identification are qualitative in nature; these are discussed in the sections dealing with hazard analysis and fault tree analysis. Quantitative aspects of hazard analysis, along with safety modeling, are discussed in the chapters dealing with hazard analysis, fault tree analysis, and the mathematics of safety.

In order to permit this text to be a self-contained entity, the elementary statistics and Boolean algebra needed for the material dealing with the mathematics of safety are included. This material is, however, generally distributed throughout the text, each portion being introduced where needed. Those sections which discuss certain probabilistic aspects of system safety, presume some knowledge of elementary calculus. This portion of the book, however, may be omitted by the reader without seriously diminishing the theme presented.

Responsibility for the effort needed to assure satisfactory safety for a given system rests upon those accountable for the design, development, manufacture, and test of the system. For moderate size and larger systems, it may be preferrable to delegate authority for carrying out this function to a systems safety analyst.

The author's experience indicates that a system safety analyst cannot fulfill his function without the collaboration of management, for they establish the extent and effectiveness of safety activities throughout the organization. In turn, management needs to know the nature and extent of the risks to be encountered and how to relate these risks to specific contractual obligations. These subjects are treated in the section dealing with achievement of safety.

An important characteristic of system safety is that it is interdisciplinary to a high degree. It must relate to various organizations and within each organization to people of different backgrounds and technical skills. Further, system safety problems range from those requiring only common sense to those requiring complex mathematics. In fulfilling safety responsibilities, it is essential that those conducting safety tasks take care to prevent any communication barrier from arising. It is hoped that this book will aslo assist in achieving that end.

Chapter 1

THE ROLE OF SYSTEM SAFETY IN SOCIETY

INTRODUCTION

Concern with safety goes back to prehistoric times, when the sole objective was survival against enemies and the elements. At that time, and for many centuries thereafter, all the hazards associated with activities people undertook were generally understood. However, as civilization progressed, many of the developments associated with progress created covert hazards whose consequences were not readily traceable to causative factors. For example, banding together into villages and cities brought about health hazards which even now are not fully controllable by techniques such as preventive medicine, sanitation standards, and sewage management. In some instances, control of a hazard by improving technology was thwarted by its reappearance in a different guise. A good illustration of this point is lead poisoning (also known as plumbism from the Latin and Saturnism from its alchemical name).

Lead has been mined and worked for millennia, its ductility, resistance to erosion, and other properties making it a most useful metal. The philosophers in ancient Rome recognized lead poisoning when they observed the mental and physical disorders in people who had a history of eating and drinking from leaden utensils. Medieval physicians recognized it when artisans working with lead complained of the "dry grippes," loss of movement in the fingers and hands. Modern physicians look for lead poisoning when presented with symptoms of central nervous system disorders, kidney damage, anemia, male sterility, and menstrual and pregnancy abnormalities in women. Despite the long history of knowledge on this subject, the possibility of lead poisoning occurring nowadays is not always remote. The use of lead as a constituent in indoor paint is controlled by law, but deaths and permanent injury occur each year as a result of lead poisoning through paint ingestion, particularly among children who live in older homes. This problem was given a modern twist in 1978 by a ruling from the Internal Revenue Service that allowed a medical deduction for part of the remedial work carried out on the home of a three-year-old lead poisoning victim.

Decorations on the outside of drinking glasses also may result in ingestion of unacceptably high levels of lead. Standards for lead content near the rim of drinking glasses were jointly established in 1979 by the Food and Drug Administration, the Environmental Protection Agency, and the Consumer Product Safety Commission. Industry provides other examples of lead poisoning. Those who work in the scrap metal industry, breaking up big ships, bridges, or steel frameworks of large buildings may be exposed to the toxic material originally used to protect the underlying metal. Workers in battery manufacturing facilities, in the paint industry, in gas stations, and in auto repair shops may also be exposed to flakes and fumes containing lead and some of its compounds. One company, Smith and Wesson, a maker of ammunition, provided an interesting solution by developing nylon jackets for lead bullets, which the company claims reduce lead oxide fumes generated about firing ranges by about 60 percent.

The problem is universal. It has been shown that the amount of lead in the polar cap has been continually increasing since the Industrial Revolution. Consider the amount of lead emitted into the earth's atmosphere through its use as an additive in gasoline. The basic physiological problem is that, while the concentration of lead in soft tissue remains relatively stable throughout life, its concentration in bone increases with age. System safety interfaces with this problem are discussed in Section 1.3.

1.2 SYSTEM SAFETY DEVELOPMENT INTO A SCIENCE

It can be considered that system safety began to develop as a science at the time people perceived their inability to have complete knowledge and control over hazards associated with everyday activities. It remains true today that people do not know about nor can they control every hazard involved in daily life. Food, clothes, drugs, transportation, hobbies, and the very air we breathe and the water we drink present daily hazards that are not fully understood.

In response to this circumstance, system safety has developed into a broad, scientific discipline with many areas of specialization, impinging upon every sphere of human activity. It is now taken for granted that government and industry will work constantly at developing new standards and improving upon existing ones to enhance safety in housing, pharmaceuticals, manufacturing, mining, and a host of other areas. Perhaps the best known of such efforts relates to automobile safety, which, with its periodic revision of safety standards, affects the design and manufacture of over 10 million vehicles annually and the lives of all who travel in them. The home is another area of concern to system safety and to the general public. Annual costs of home accidents in the United States have consistently risen and were estimated to be about $1.7 billion in 1978, costs which cannot accurately reflect the value of injuries and lost lives. (The term *value* is formally defined in Section 2.10.)

Another area in which safety is of major concern in the everyday lives of millions of people is job safety. In the early 1970s increased attention was focused on this subject as a consequence of major federal legislation. Almost everyone agreed that sound legislation was needed to ensure the safety and health of America's 100 million workers, for government statistics indicate that more than 14,500 workers are killed on the job each year and that 2.2 million are disabled. In 1970 President Nixon signed into law the Occupational Safety and Health Act (OSHA), which established occupational health and safety standards for employees of firms engaged in interstate commerce. At the federal level, the bill affects 55 million workers. In addition to issuing health and safety standards, the bill provides for a board to adjudicate disputes arising out of the enforcement of the law and establishes a national institute to conduct research on job safety.

1.3 THE ESSENCE OF SYSTEM SAFETY

The essence of the system safety discipline and its usefulness to society can be described through further examination of the hazards associated with lead and by examining the positions espoused by industry, government, and labor on this subject.

The amount of lead required to cause injury has not been precisely established. Early in this century 500 micrograms of lead per cubic meter of air (500 $\mu g/m^3$) was considered to be acceptable. This was subsequently lowered to 200 $\mu g/m^3$, and a level of 100 $\mu g/m^3$ is now proposed by the federal government. The arguments for and against the proposed standard and about the methods proposed for its enforcement are typical of the problems that confront the safety analyst. Associated problem areas, such as those concerned with rate retention and equal opportunity, embroil the safety specialist in system considerations. Determining the amount of lead that does cause injury does not lend itself to a unique solution. Rather, only a compromise solution, ideally a compromise that is also an optimum solution, can be achieved. The elements of such a compromise are discussed next.

1.3.1 Merit of Proposed Standard

The standard of 100 $\mu g/m^3$ proposed by OSHA was based upon assessments that

1. Clear-cut clinical symptoms of lead intoxication appear in workers whose blood levels are approximately 80 μg lead per 100 g blood.
2. . . . if air lead levels of 100 micrograms are maintained, the maximum upper blood level of workers should remain below 60 micrograms.

Labor objections to the proposed standard were based upon arguments that

1. Airborne lead exposure should be restricted to 50 micrograms per cubic meter of air [since] . . . the proposed restriction to 100 micrograms air . . . is not enough of an improvement.
2. The only exposure condition which can be considered safe . . . results in no significant increase in body burden of lead over a preemployment level, or the average of the general population.

The difference in cost between implementing the current standard (200 μg/m³), the proposed standard (100 μg/m³), and the one suggested by labor (50 μg/m³) involves questions such as the following.

1. How safe is safe?
2. What is the value of 50 μg/m³ as compared to 100 μg/m³?
3. Is exposure at 200 μg/m³ for 10 days equal to exposure to 100 μg/m³ for 20 days?
4. Is exposure to 100 μg/m³ for 15 years no more harmful than exposure to 50 μg/m³ for 30 days?

Answers to these questions depend primarily on scientific investigation. The overall problem, however, cannot be divorced from political and economic questions such as, Does money spent to achieve a lower level place a drag on the United States economy or lower the competitive position of United States workers versus those in other countries?

1.3.2 Methods for Measuring and Controlling Exposure

In addition to disagreements about acceptable levels of lead exposure, there is disagreement about methods for measuring and controlling exposure. Labor argued for environmental monitoring, that is, controls over the amount of lead in the atmosphere, while industry favored biological monitoring in the form of blood and urine testing. The former is achieved by engineering controls, designs intended to prevent lead from entering the atmosphere. The latter is achieved by respirators that prevent lead in the atmosphere from being inhaled. Resolution of this disagreement involves technical and economic considerations.

A factor that is almost entirely in the economic sphere concerns rate retention. Rate retention means that workers who are found to have elevated levels of lead in their blood and who are moved to safer but lower-paying positions remain at their former pay scale. A rate retention clause was incorporated in the 1977 Mine Safety and Health Act and threatened to become an issue during hearings on coke oven emission standards. The lead standard appears, however, to be the primary battleground for rate retention because of the large number of employees involved (835,000) and because lead workers have traditionally been moved around when they evidenced elevated blood lead levels.

1.3.3 Lead Poisoning and Equal Opportunity

Lead is one of the toxic substances, mutagens, gametoxins, and teratogens that adversely affect reproduction and cause sexual dysfunction. The need to prevent these adverse effects is becoming more urgent because women are entering the labor force at a rate of almost 2 million per year. Most are in the childbearing ages of 16 to 44, and approximately 1 million prenatal infants are in American workplaces each year. Some companies prohibit women of childbearing age from working in areas where they could be exposed to lead (and other toxic substances). Such policies can create tension between the objectives of making the workplace safe for all and providing equal employment opportunity. (In some instances women have opted for voluntary sterilization rather than accept transfers to lower-paying positions. To further complicate this emotional issue, it was subsequently determined in some of the cases of sterilization that the chemicals in question were not toxic.) Exclusionary policies have not satisfied men either, for reproductive capacity in males can also be affected by lead.

1.4 CURRENT STATUS OF SYSTEM SAFETY

System safety comprises two parts. One is technical and is concerned with identifying hazards and their effects and determining the likelihood of their occurrence. The other involves less tangible considerations, such as ethics and value judgments. Both are inextricably entwined and often lend themselves to resolution only by political process. On occasion, there is an interface between equipment and human behavior that influences system safety in still a different fashion. Humans operate systems, such as automobiles, trains, aircraft, and reactors. Some of these systems have automatic sensing and controls to deal with the onset of accidents, others have few such devices and rely on human intervention for accident control. In any case, the number of possible accident scenarios prevents computers from having total control, so that a human-machine interaction occurs after the onset of an accident.

While people are innovative and resourceful, capable of minimizing potential accident consequences, they are also fallible, capable of aggravating and compounding a minor problem. The chain of events leading to an accident may, in fact, be caused by human behavior. Resolution of the human factor in unsafe conditions is a technical problem, one in which great progress has been made. Quantification of human behavior, however, will probably never equal that achieved for predicting hardware failure. (Estimation of human failure rate is discussed at greater length in Section 8.11.)

As in the case of so many other bodies of knowledge, the rate of progress in system safety has been greatest during recent times. The growth of system safety during the last few decades is particularly evident in the domain of astronautics. Final preparation of a modern craft for launching astronauts into space on a minimum risk basis includes a lengthy period of pad checkout using a comprehensive array of computers and instruments. Simulated countdowns, test fueling, and emptying of tanks, with electronic sensors sniffing for leaks, are conducted. At the proper moment, a computer initiates ignition of the engines and sensors measure whether thrust is building properly and, if necessary, order and execute a shutdown. Such a scene is in striking contrast to that day in 1926 when the American rocket pioneer Dr. Robert H. Goddard launched the world's first liquid-fueled rocket from a Massachusetts cow pasture. At that time an assistant started the engine with a blowtorch tied to a long stick and ran for cover.

1.5 SYSTEM SAFETY IN THE FUTURE

Until recently, the most intensive system safety efforts have been focused on two general classes of activity. The first, engaged in by relatively few people but of great interest to the general public, relates to human travel to new and unfamiliar environments—into outer space, through the atmosphere at great heights and speeds, and into the depths of the ocean. Efforts in these areas are more strongly oriented toward the technical aspects of system safety. The second, more recent in character, relates to the prevention of events that are potentially catastrophic to large numbers of people. Examples include core meltdown in a nuclear electric-generating plant, crash of a large passenger aircraft, and long-term effects of chemicals and pollutants breathed and ingested. Such problems require both technical safety considerations and system considerations.

For the future, system safety needs to apply the techniques developed for esoteric problems such as space flight and inadvertent nuclear explosion to those areas of everyday life that are most profound in terms of the numbers of deaths and injuries that occur. The development of new life-saving, labor-saving, and life-enriching devices is desirable. However, the introduction of injuries and long-term fatalities as a by-product of such innovation can cause restraining forces to be brought upon developers and manufacturers by the public and the government. Developers may respond in turn by diminishing their efforts to bring about such devices.

Solutions to these problems cannot be achieved without the application of system safety methodology because of the ever increasing rate at which technology is advancing and because of the increasing requirement to incorporate economic and ethical considerations in the solutions.

Additional Reading

Air/Water Pollution Report. Weekly. Silver Spring, Md.: Business Publications, Inc.

Armstrong, R. The passion that rules Ralph Nader. Fortune, May 1971, pp. 144–147.

Chisolm, J. J., Jr. Lead poisoning. *Scientific American*, February 1971, pp. 15–23.

Cooney, J. E. Hospital hazards. *Wall Street Journal*, November 16, 1970, p. 1.

Engel, P. G. Controversy over the lead standard moves into the federal courts. *Occupational Hazards*, April 1979, pp. 53–56.

Lewis, H. W. The safety of fission reactors. *Scientific American*, March 1980, pp. 53–65.

National Safety Council. *National Safety News; Safety News Letter* (Marine Section, Electric Section, Hospital Section, and Annual Industrial News Letter); *Family Safety*; and *Traffic Safety*. Chicago: The Council.

Nernec, M. M. Warning: This job may be dangerous to your offspring. *Occupational Hazards*, April 1979, pp. 37–40.

Spivak, J. Shape of the future. *Wall Street Journal*, January 6, 1967, pp. 1, 14.

Chapter 2

THE LANGUAGE OF SYSTEM SAFETY

2.1 DEFINITION OF SAFETY

Etymologically, the word *safe* is traceable to several sources. The Latin *salvus* translates into safe, whole, or healthy and is akin to *salus*, which may be translated as health or safety. The derivation from the Greek relates to the word *holos*, which means complete or entire; and the Sanskrit word *sarva* means unharmed or entire. The process by which these roots were transformed into the modern English adjective *safe* becomes evident through an examination of the old French variations, *salf, sauf, sof,* and *sal,* and the variations used in Middle English, *sauf, saf,* and *save.*

Webster's New International Dictionary (unabridged), defines *safe* as "free from harm, injury, or risk; no longer threatened by danger or injury; secure from threat of danger, harm, or loss." The noun *safety* is defined by Webster's as "the condition of being safe; freedom from exposure to danger; exemption from hurt, injury, or loss."

There are, therefore, a variety of definitions for safety, any one of which can be used as a satisfactory starting point for system safety considerations. The definition used in this text, in our opinion, reflects the true essence of the term.

> *Safety* is freedom from conditions that can cause injury, illness, or death to personnel or damage to or loss of equipment or property, or environmental harm.

Some of the qualities that make this definition of safety suitable for use in applying the concepts of system safety are

1. It is general, in that it relates both to people and to things and in each instance to partial or total loss. Nevertheless, the definition also lends itself to restricted versions for application to those instances where limitations are preferable. The definition could, for example, be limited

to "injury or death" or to "damage to equipment." Such restrictions are subsets of the general case.

2. The definition is qualitative rather than quantitative. This aspect is considered a generalizing rather than a restricting factor because system safety studies, as will be shown in subsequent chapters, logically proceed from a qualitative, general form to a more detailed, quantitative form.

3. The definition permits a natural interface to exist between safety and other disciplines with which it is closely allied. One major group of allied disciplines, often referred to as the "ilities," are system effectiveness, reliability, maintainability, quality assurance, and human factors. The nature of these interfaces and their significance to safety are the subject matter of the next chapter.

2.2 QUANTIFICATION OF SAFETY

The definition established for safety is not readily quantifiable in its present form. Quantification, however, is necessary if a consistent, reliable estimate is to be obtained about the "safeness" of undertaking a given task. The notion of quantification is introduced at this early stage because of its relative importance in the discipline of system safety.

It might be noted that everyone is called upon to make quantitative assessments, at least at the subconscious level, for a variety of apparently routine tasks. Consider, for example, the process that takes place in changing lanes while driving. The conventional procedure is one that obviously begins with an assessment of the circumstances. In most cases the resulting quantification is no more precise than "zero" or "one," implying, respectively, do not change lanes at this time or it is safe to change lanes provided a greater distance or a greater speed is established prior to initiating the maneuver. It is reasonable to presume that a considerable number of accidents occurring during lane changing result from errors drivers make in assessing the safeness of undertaking the task. Adding microcomputers to automobiles with inputs from various sensing devices, as proposed in some highway safety studies, could permit a more precise, quantitative assessment of driving tasks such as changing lanes and, consequently, could decrease the number of accidents.

In the safety domain, quantification is accomplished in one of two ways. In one, a probability is assigned to each member of the set of events that make up the task, and the several probabilities are then combined into one probability for the system. This method of quantification is suitable when the effects of an event's occurrence are known but the likelihood of its occurrence is not known. A procedure carrying out such quantification is discussed in Chapter 9. The other type of quantification is concerned with establishing the consequences of an event in terms of the intensity of its effects. For example, almost everyone is aware of the danger of electrocution from contact with electrical energy. Efforts to quantify the effects of 60-Hz alternating current on humans has been under way for some time now. More recently, efforts at quantification have also been oriented to concern that fibrillation (discoordinating heart action) or respiratory paralysis can be caused by currents as low as 10 milliamperes.

2.3 HAZARD CONSIDERATIONS

The antithesis of safety, and the "villain" of this text, is the hazard. This "villain" is ubiquitous in real life because there are inherent hazards associated with each and every human activity, even sleeping. As in the case of safety, there are a variety of definitions that may be considered to be acceptable. The one we have chosen to define *hazard* is

> An existing or a potential condition whose occurrence can result in a mishap.

In turn, *mishap* is defined as

> An unplanned event or series of events that results in injury, illness, or death to personnel, damaged to or loss of equipment or property, or environmental harm.

Like the definition of safety, the definition of hazard is also qualitative in nature. Quantification in this case involves

1. Establishing a measure for the term "potential"
2. Assessing the seriousness of the injury or damage

The term *risk* is sometimes used to describe the estimated probabilities of harm likely to be associated with conducting an activity. Without additional clarification, however, it is not clear whether risk refers to probability of occurrence or seriousness of consequence. Some of the techniques for quantifying probability of occurrence are discussed in Chapters 8 and 9.

Quantification of the seriousness of the injury or damage is the more complex of the two factors because relative judgment is required. Clearly, those hazards which affect human survivability have the highest weighting factor. If, however, survivability is not in question, then it is necessary to determine whether continuation of the activity is sufficiently desirable to merit acceptance of the consequences resulting from the existing hazards. Typical categories that need to be considered when survivability is not in question include

1. Decrease in potential life span
2. Diminution of the quality of life

3. Nature of potential injury
4. Decrease or loss of operational capability
5. Damage to or loss of property

The seriousness of these five categories is relative, depending upon the circumstances surrounding the event. There is, for example, a risk of shortening one's life span as a result of exposure to X rays. Such a risk is not considered unreasonable, however, if the X rays are used to assist in setting a broken bone. By way of contrast, fluoroscopy was used about 50 years ago in shoe stores as an aid in fitting shoes for children.

If it can be established a priori that the existing hazards relate only to property damage, then the willingness to continue with a project becomes an exercise in economics. There may be a complicating factor in that time and perhaps other variables may need to be transformed into an equivalent dollar value. For example, the replacement cost of a defective component is often small in relation to the expense incurred in carrying out the repair or to the cost incurred because the equipment is temporarily out of service.

2.4 HAZARD CLASSIFICATION

There have been a variety of attempts to classify hazards as a function of resulting severity. Inasmuch as the range of severity remains fixed, varying between inconsequential or minimum property damage and death, the classifications proposed differ mainly in the number of categories established. Since it is often difficult to distinguish between the relative importance of the severity of nonfatal hazards, no useful purpose is served by having a large number of categories. Evidence of this difficulty is demonstrated by the inability to standardize the relative importance of identical injuries to different persons in automobile accidents, despite the unfortunate fact that there is a large sample available for analytical purposes.

This text recommends a mechanism for classifying hazards that makes use of the following four categories:

Class IV—*Safe*. Conditions such that human error, deficiency or inadequacy of design, or equipment malfunction will not result in personnel injury or equipment damage.
Class III—*Marginal*. Conditions such that human error, deficiency or inadequacy of design, or equipment malfunction will degrade system performance or damage equipment but counteraction or control can be undertaken such that serious injury or significant damage will not occur.
Class II—*Catastrophic*. Conditions such that human error, deficiency or inadequacy of design, or equipment malfunction will severely degrade

system performance and cause subsequent system loss and/or cause death or serious, irreversible injuries to personnel.

Class I—*Critical.*[?] Conditions such that human error, deficiency or inadequacy of design, or equipment malfunction will cause personnel injury and/or serious equipment damage or will result in a hazard requiring immediate corrective action for personnel or system survival.

This categorization of hazards is considered to have a universal character in that each of the four classes relates to injury, equipment damage, or both. In addition, injury is defined in its most general sense including illness, throughout this text. There could be some difficulty in selecting between two adjacent classes when attempting to establish a precise classification for a given hazard, even with as few as four categories. Two considerations are offered as aids in determining the exact hazard classification, one for injury and one for equipment damage. In the case of injuries, greater importance is usually applicable to those that are residual (irreversible) than to those that are transient (reversible). These terms are defined as follows:

A *transient injury* (or illness) is one from which recovery is effected with no resultant loss of functional capability or shortening of life span.

A *residual injury* is one that is not transient.

Loss of a limb, for example, is residual. A broken bone in a limb is transient if nominal ability to use that limb is restored; otherwise, it is residual.

In the case of equipment damage, classification of a hazard can be more easily determined if the hazard's effect upon the system is known. For example, damage to an aircraft hydraulic system that is used to raise and lower the landing gear may be considered to be a Class III hazard if the failure prevents the gear from retracting and the gear remains in a locked position. The identical failure could be Class I if it occurs while the gear is in a retracted position. However, the latter circumstance reverts from Class I to Class III hazard if a mechanical backup system exists which can be used for lowering and locking the landing gear in place. Cost or time considerations may also be used in establishing hazard classifications based on their importance to the system or its use.

One additional thought on the subject of classifying hazards relates to the notion of "negative" hazard. Such a hazard might be illustrated by a transient injury from which recovery is effected with a positive gain of functional capability. Consider, for example, a bacterial illness that leaves no residual negative effect and imparts a lasting immunity.

These four hazard classes, I to IV, are summarized in tabular fashion in Table 2-1, showing the separation between injury and equipment damage (or environmental harm). The presumed occurrence of just one personnel injury or equipment damage may be sufficient to establish the classification.

Table 2-1
Hazard Classification

Class	Hazard	Equipment Damage	Injury
IV	Safe	None	None
III	Marginal	Minor	None or minor reversible injury
II	Critical	Substantial	Major transient injury or minor irreversible injury
I	Catastrophic	System loss	Irreversible injury or death

2.5 THE HAZARD VECTOR

It has been shown that the hazard level (criticality) associated with under-taking a given task is a function of (1) the severity of the hazard's effects (i.e., its hazard class, C) and (2) the probability that the hazard will occur (P).
That is,

$$HL = f(C_i, P_i) \qquad (2.1)$$

where

HL = the hazard level
C_i = the weighting factor associated with the ith hazard, based on the four classes established
P_i = the probability that the ith hazard will occur

The pair of values (C_i, P_i) that is associated with the ith hazard defines the hazard vector, H_i. There is an analogy between the hazard vector and the conventionally defined two-dimensional vector. In the analogy, C is roughly equivalent to direction and P to magnitude. As in the case of conventional vectors, $H_i = H_j$ when $C_i = C_j$ and $P_i = P_j$. The analogy also holds in relation to the addition of the hazard vectors, including the validity of the commutative, associative, and distributive laws.

It might be inferred from Equation 2.1 that an appropriate strategy to follow is one that minimizes the hazard level. However, such a strategy is valid only under circumstances when safety is independent of other disciplines, a situation that is not generally true.

2.6 DEFINITION OF SYSTEM

System safety relates and combines the characteristics and implications of safety to those of systems. Consequently, it is appropriate to discuss systems and how they interface with safety. In the general sense, a *system* may be defined as

A device, scheme, or procedure which behaves in accordance with some description, its function being to operate on information and/or energy and/or matter in some time reference in order to yield information and/or energy and/or matter.

A definition oriented more closely with system safety is

A composite, at any level of complexity, of personnel, materials, tools, equipment facilities, environment, and software. The elements of this composite entity are used together in the intended operational or support environment to perform a given task to achieve a specific production, or to support a mission requirement.

These definitions place no restriction on the size or complexity of the device, scheme, or procedure under consideration. Moderate-sized or large systems usually comprise some composite of operational and support equipment, personnel, facilities, and software that are used as an entity to perform or support a specified role. Examples of large systems include conventional transportation equipment such as trains; uncommon transportation devices such as deep submersibles and spacecraft; and maintenance systems such as those used for commercial airlines, military weapon systems, and hospitals.

Any one of these examples may be expanded or contracted in size or scope of function and still fall within the definition. A train system, for example, may be expanded into a network made up of several train systems, and a hospital system may be expanded to include its relationship to medical schools, nursing schools, the drug industry, medical insurance, and other related equipments and procedures. Conversely, a portion of a systems may, by itself, be considered a complete system. Examples of this latter case include a single jet engine, a locomotive gear train, a hospital operating theater, one human being, or, subdividing a human further, the circulatory or skeletal system.

The operational role for the primary function performed by a system is referred to as its *mission*. (This role may be invariant, or it may vary as a function of time as defined in section 4.2.1.) Additional assistance in classifying hazards is obtained by considering the effect the hazard's occurrence could have upon the system mission. A portion of the mission may be sufficient for use in an analysis of the relationship between hazard occurrence and mission effectiveness.

2.7 DESCRIPTION OF SYSTEM

A system may be described by specifying all of the following:

1. Its inputs and outputs, $E_i(t)$ and $E_o(t)$, respectively, as a function of time

Figure 2-1
Schematic Representation of a System

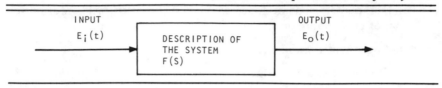

2. All the possible conditions (states) of the system, that is, the system phase space
3. A model relating inputs, outputs, and system states as a function of time, $F(S)$

$F(S)$ and its relation to E_i and E_o are shown schematically in Figure 2.1. In general, the operations performed by a system and, consequently, its capability to carry out a specified mission depend upon the condition (the internal state) of the system. It is necessary, therefore, to consider the total dynamics of the system when describing its phase space and its model, $F(S)$. In addition, the model for $F(S)$ needs to describe both the steady and transient states of the system, where these terms are defined as follows:

A *dynamic system* is said to be in a steady state condition when the variables describing its behavior are either invariant with time or are periodic functions of time (or sections thereof).

A dynamic system is in a transient (unsteady) state when it is not in a steady state. This definition for transience of a system agrees with the one given in Section 2.4 for transience of injury or illness.

In some systems certain of the inputs or phase states are incorporated solely because it is known that certain specified hazards will occur, or because they are presumed to be extremely likely to occur. These inputs or phase states are known as contingency plans or operations. The use of redundancy in the design of a system is one type of contingency plan.

2.8 DEFINITION OF SYSTEM SAFETY

A definition of system safety can be derived by restricting the inputs, outputs, and system model description to just hazard considerations. To illustrate this process, consider a passenger airliner and its crew as a system, and presume that a system mission has been defined in terms of transporting various mixes of passengers and cargo as a function of distance. Some considerations of the elements of E_i, $F(S)$, and E_o of this system include the following:

1. System inputs, E_i, include, among hundreds of others, a description of weather conditions, maintenance and overhaul activities, qualification and training requirements of the crew members, and contingency operational plans.
2. The system model, $F(S)$, includes aspects such as the rate of fuel consumption as a function of altitude and speed; yaw, pitch, and roll coupling characteristics as a function of wind gusting; and variations in the response of the tires and braking mechanism as a function of the aircraft's ground speed and differing runway conditions.
3. The system output, E_o, when considered to be the resultant of hazardous inputs and the response of the system to those inputs, can take on one of two forms. In one, no hazard occurs, or any hazards that occur are adequately controlled. In this instance E_o represents safe conditions. In the other case, one or more hazards do occur and at least one is not adequately controlled. E_o represents the unsafe circumstance under these conditions.

In a general sense it can be said that system safety is concerned with providing procedures and equipment for preserving the integrity of the system, its occupants, and other contents over the range of environmental and operational conditions that can reasonably be expected to occur during the mission.

A formal definition of system safety needs to relate to two fundamental aspects, control and reasonableness. The first aspect, associated with the identification and control of hazards for preserving the integrity of the equipment and occupants so that the mission may be completed, relates to the "safety" part of system safety. The other aspect, arising from the phrase "reasonably be expected to occur," is associated with the "system" part of system safety. The task of determining what is reasonable or acceptable for a given set of conditions is necessary because the ideal situation, absolute safety, can only be established as a goal but can never be achieved in practice. (Later in this chapter it is shown that absolute safety is rarely practicable even as a goal.) The following definition of system safety brings the aspects of control and reasonableness together.

> *System safety* is an optimum degree of safety, established within the constraints of operational effectiveness, time, and cost, and other applicable interfaces to safety, that is achievable throughout all phases of the system life cycle.

In this definition the notion of reasonableness, which is qualitative and subject to varying interpretations, is transformed into the more formal concept of optimization, which is quantifiable. The definition is not meant to imply that one unique optimum is appropriate or desirable as a definition of system safety for the life of the system, although this possibility is not unacceptable. Rather, the definition establishes a requirement that systems analysis techniques include a quantification of safety over the entire life of the system based on all facets of the system, E_i, $F(S)$, and E_o. *Optimization*, which is elaborated on in the next section, is the essence of system safety and, as described in this text, is primarily derived from the domain of systems analysis. It may be defined as

> The application of mathematics and simulation techniques for identification, examination, and calibration of the interaction between and among the elements of a system.

The methodology of systems analysis, initially advanced by the aerospace industry, is now seen more universally as a logical approach to problem solving. Many of society's social needs, transportation, ecological control, and educational motivation of the hard-core disadvantaged are all examples of problems whose solutions are being attempted by systems analysis technology. Its relation to system safety is an application of systems analysis to one specific area and is treated at some length in succeeding pages.

2.9 OPTIMIZING SYSTEM SAFETY

Achieving an "optimum degree of safety" requires the application of scientific knowledge and methodology to assist in making a choice among the various alternatives available for arriving at a chosen objective. Alternatives refer to the various configurations and options possible for the system description, $F(S)$, and for the inputs, E_i. In practice, these alternatives are to be found within the domains of engineering, reliability, maintainability, personnel training, design, and any other discipline that significantly influences or interfaces with safety. Alternatives that could be selected in an effort to produce an optimum degree of safety include system configurations that are oriented toward

1. Minimum complexity. There is a saying in the automobile industry to the effect that problems cannot arise in equipment left off the machine.
2. Placing minimum demands on human skills for operation or maintenance.
3. Assuring that failure of any one component cannot lead to incapacitation of the system or to fatality.
4. Providing an indication of those components that have become degraded and, consequently, are likely to fail.

An evaluation of alternatives cannot be made without relating to the system outputs, that is, to the system objectives, E_o. However, it is difficult to attempt to quantify system objectives when they are not a simple listing of desirable goals or events. In such a circumstance, the set of system objectives may be visualized as a preference surface which relates these objectives to their "value" or "utility." The need to solve this type of problem almost always exists in the domain of system safety when people are involved, because it is considered undesirable in our culture to evaluate human life in terms of inanimate equipment or money. Consequently, E_o and $F(S)$ are often based upon a system requirement that people not be harmed. Similarly, the notion that risks may be intentionally taken as part of the operation of a system, which includes a schedule of compensation for any injuries or fatalities that may occur, is equally undesirable in our culture.

If system considerations do not involve human safety, there is less hesitancy in permitting E_i to range over the domain of all possibilities in order to establish an optimum for each possible E_o. Suppose, for example, that a manufacturer desires to compare the various alternatives available in a system for delivery of spare parts to customers. Suppose further that the system objective, E_o, is to maximize profits. Possible approaches as system inputs, E_i, include

1. Maximizing expected profits, presuming they can be projected in the marketplace.
2. Maximizing revenues for a given cost, or minimizing the cost of generating a given volume of revenue.
3. Maximizing the volume of deliveries within a specified time period for a given cost, or minimizing costs or achieving a specified type and number of deliveries.

The firm now needs to choose one strategy from among the many available for implementing the spare parts delivery system so that the stated objective will be achieved. The validity and, consequently, the usefulness of attempting to calculate a quantitative solution for this problem depends on the shrewdness with which the various alternatives are selected. Presuming that the list of alternatives presented above is complete and valid, it can be perceived that the developer of the system need not hesitate in establishing quantitative measures for the gain to be achieved in pursuing each alternative and in establishing an analogous measure of the risk associated with each. If, however, one of the alternatives includes a likely possibility of serious injury or fatality, then it would be expected that this alternative would be eliminated from the set of alternatives available for selection by the system developer.

Optimizing a system objective that is expressed as a preference surface introduces the notion that safety may have a value that varies from person to person and that it may even be variable for the same individual under different circumstances. (This aspect of subjective optimization, apparently a contradictory phrase, is discussed in the next section.) In addition, use of a preference surface requires that some consideration be given to the value an individual presumes will be gained or lost as a result of undertaking a task with a given risk in comparison to the value established by society for that undertaking. This aspect of system safety in which individual and collective opinions differ is also a subjective process that needs to be quantified; it is discussed in Section 2.11.

2.10 THE VALUE OF SAFETY

The fundamental problem that prevents value from being simply defined for use in the safety domain is that both tangible and intangible considerations need to be applied as system constraints. A further complication exists because any given tangible or intangible figure of merit used for value can be assessed simultaneously in an absolute and in a relative fashion. If a separation is permitted between the two, the definition becomes more manageable. Absolute value can be defined in terms of cost (price) as follows:

> *Cost value* is equal to the amount of money needed to pay for labor, materials, and overhead that are required to produce a given item or to establish an explicit system output, E_o.

Relative value is subjective. It could be oriented ethically, concerned with the need to establish values on human life, or it could be profit-oriented, utilizing product safety as a competitive advantage. In either instance, relative value may be defined in terms of "use" or "esteem":

> *Use value* relates to the properties and qualities of an item or system output that permit a task, work, or service to be performed.

> *Esteem value* relates to the characteristics of an item or a system output that make the system desirable or attractive and, consequently, valuable.

Use value is relatively objective, and rules can be formulated for its calculation; esteem value is more subjective. Both can have instantaneous cost value, but no steady-state, constant value can be established for either. A definition of value that combines both absolute and relative characteristics may be given in terms of "exchange":

> *Exchange value* is determined by those intrinsic properties of an item or system output, E_o, that enable it to be exchanged or traded for some other item or output.

Exchange value has some of the qualities of relative value, but its use forces some consideration of absolute value in the optimizing process. One formal procedure for transforming relative value into absolute value is through the application of game theory, which is introduced in the next chapter. Ideally, no activity should be undertaken without some quantitative assessment of the risks involved, which cannot be accomplished in all instances without some transformation of relative safety values into absolute ones. This notion is generalized further in the next section.

2.11 PHILOSOPHY OF RISK ACCEPTANCE

There can be a significant difference between the measure of a safety level obtained through the process of optimizing system outputs, E_o, and the measure obtained through subjective means. Optimum safety is mathematical in nature and is characterized by an attempt to determine all possible alternatives. Since all alternatives cannot practicably be used in the quantification process, it is necessary to select those that have the greatest influence upon E_o.

Subjective safety level is more philosophical in nature, and is based on

1. Some notion of the risks that must be undertaken in attempting to design, manufacture, or use a system.
2. An understanding of the adverse effects that can occur.
3. An indication of the likelihood of occurrence of these adverse effects.
4. The willingness of the individuals participating in the venture to accept the risks, which in this instance are based on a subjective assessment of the value of the payoff.
5. A measure of the willingness of society to permit the individuals involved in the venture to accept the risks.

If the risks involved and their effects and probability of occurrence are known absolutely, then acceptance of the undertaking by the individual and by society can be assessed by evaluating the potential social or economic benefit to be derived by society and/or the individual. One additional factor—the individual's role in society—is tacitly taken into consideration by society when it permits individuals to assume certain risks. For example, initial experiments conducted on humans to determine the effects of new drugs are often carried out on volunteer prisoners or as a heroic effort upon the critically ill.

A code drawn up in 1964 by the World Medical Association distinguishes between two types of human medical experiments. In the more common of the two, a method of treatment is attempted on an ailing person with the primary intent of helping the patient, for example, the use of an experimental antitumor drug to attempt a remission in advanced cancer. The other type of experiment is conducted for the purpose of gaining new data but does not provide any direct or immediate benefit to the subject. A well-known example is Dr. Walter Reed's attempt to determine the vector for yellow fever. The subjects in his experiment were volunteers who were legally competent, aware of the risk undertaken, and able to withdraw from the experiment at any time.

The ethics of risk acceptance are focused most sharply and consequently cause great controversy when the subjects are incapable of understanding the risk to which they are exposed; that is particularly true when young children are used as subjects. One such controversy flared up in 1971 concerning experiments that had been taking place for 10 years among mentally retarded children at the Willowbook State School in New York. The purpose of the experiments was to find the cause and, it was hoped, a technique for preventing serum hepatitis. The controversy at Willowbrook centered on doubts about the benefits to be derived by the subjects and the relevancy of parental permission under such circumstances. Yet it must be noted that the experiments at Willowbrook contained parallels and precedents. One of the earliest took place in 1796 when Dr. Edward Jenner inoculated eight-year-old James Phipps with material from a young lady's cowpox sore in an unsuccessful attempt to infect the youngster with smallpox. More recently, the Salk polio vaccine was tested on institutionalized children in its early development with results that subsequently prevented death and disability in thousands of other children. Perhaps the most relevant precedent is the development of measles vaccine at Willowbrook under conditions almost identical to those prevailing during the hepatitis experimentation.

2.12 REGULATED vs. UNREGULATED RISKS

In attempting to establish criteria for risk acceptance, a distinction needs to be made between (1) unregulated risks taken freely by individuals or by large segments of the population for economic reward or personal gratification and (2) risks that for logical or other reasons are regulated by government or private agencies. The former category includes activities such as engaging in dangerous sports and accepting obesity as a life-style. The latter category includes, by definition, all those tasks and activities that are regulated, such as speed limits on highways. There are a sufficient number of regulated activities to obviate the need for additional illustrations. The interface between the two categories, controlled and uncontrolled, is quite dynamic. The use of marijuana, for example, is an activity whose need for control is controversial, with strong positions pro and con. There are some activities for which agreement exists about a need for control but for which complete control is not feasible. The desire by society to prevent individuals from driving while intoxicated or under the influence of drugs is one example. In other cases the interface is blurred because of disagreements among regulators: cigarette smoking is under attack by a variety of federal, state, and local agencies while still other agencies provide subsidies to tobacco growers. There are even some safety interfaces that involve interpretations of the

United States Constitution, such as the issue of television commercials that exhort children to eat sweetened cereals and candies. On one hand are individuals concerned about a potential health hazard and on the other hand are people concerned about limitations on freedom of speech.

One characteristic common to much discussion on safety regulations is that arguments waged pro and con are often based upon emotions rather than upon systems analysis or optimization. The steps that take place when a system is evolving from an uncontrolled environment to a controlled one is shown in Figure 2-2. First attempts at control by regulation are often made without applying systems technology. In time, awareness of all alternatives and their effects upon the system output, E_o, are learned, and a systems approach is implemented.

The interval of time between the unregulated approach and the regulated systems approach shown in Figure 2-2 can be quite long; the evolutionary process that takes place during this time is governed almost entirely by political processes. As a general rule, systems whose inner workings are less understood by laypeople are likely to achieve a systems approach more rapidly. In the areas of aircraft safety and safety in medical treatment, for example, the mechanisms by which safety is accomplished are more likely to be debated by experts than by laypeople. Controls for enhancing safety aspects of the air we breathe and the water we drink, on the other hand, are subjects that permit debate among experts and nonexperts alike.

The change from an unregulated to a regulated approach for hazards associated with air and water illustrates how this evolutionary process is affected when political and economic considerations enter the debate. This is not to suggest that such considerations are inappropriate. They are, rather, a necessary and useful component of the problem-solving process in a free society, an aspect amplified further in the next section. When efforts for ecological control of air and water were first attempted, different system outputs were specified by similar and different levels of governmental

Figure 2-2
Evolution of Hazard Control

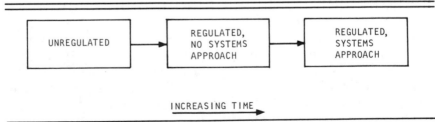

agencies, located contiguously and far apart, and based upon differing payoffs and strategies. Some attempts have been made to establish uniformity by regulations formulated at the federal level. If nothing else, these federal regulations have revealed the diversity of opinion existing at local levels. Achieving federal standards for some water supplies, for example, requires expenditures claimed not to be available by local communities. In some instances businesses have indicated a preference for closing facilities rather than spending the money needed to achieve required air standards. In turn, this has shown a willingness on the part of some people to risk living with lower air quality but with greater economic security. In other instances, objections were made to the federal standards for opposite reasons—because it was felt their acceptance would degrade existing air quality.

2.13 THE NEED FOR RISK ACCEPTANCE

In general, the search for optimum safety, an inherent component of system safety, necessitates undertaking risks. It can be expected, for example, that some disagreement will exist on the exact degree of safety required on most projects. Nor is it always an acceptable strategy to be cautious by erring on the safe side. The extent and nature of the risks needed to optimize safety for systems that affect many people are usually the subject of considerable debate, incorporating ethical, personal, and economic considerations. Economic considerations are, in fact, assuming a greater importance in such debates, perhaps because economics is a universal common denominator. As a case in point, the U.S. Supreme Court agreed in February 1979 to consider whether the government must weigh costs against the number of lives likely to be saved by a job-safety standard before forcing industry to comply. A lower court had struck down an OSHA standard limiting worker exposure to benzene when the court agreed with industry claims that OSHA had not properly taken cost-benefit factors into account in establishing a benzene standard in 1978. In turn, the government and union groups sought high court review of this lower court ruling, claiming that the ruling could undermine worker protection against benzene and other toxic substances.

The risks associated with any hazard cannot always be reduced for everyone. Attempts to do just that result in risk displacement rather than risk elimination. A ban on saccharin may be effective in controlling one carcinogen, but it could result in an increase in obesity, thereby increasing the risk of heart attacks. In addition, such a ban could make it more difficult for some diabetics to avoid sugar, and its concomitant risks, in their diets. Nitrates in foods may also cause cancer, but their elimination could increase the number of deaths caused by botulism. The Environmental Protection Agency is considering, under the Toxic Substances Act, banning the use of creosote because of its potential risk to humans. Particularly in warmer climates where wood-eating insects thrive, such a ban could result in termite-infested railroad ties with a resultant increase in railroad accidents.

It is a sign of our times that a ruling made by one government agency to reduce risks or accidents is opposed by another agency that feels the ruling would increase risks and accidents. This notion leads to the question "Should people be allowed to take risks?" A corollary question might be "Should cars and factories, for example, be engineered so that risks can't be taken?" In turn this leads to the question "Can this be done?" Reliance on expert opinion to answer these questions, a conventional way to resolve differences about the effects of technology, is not always helpful. Rather, it is also a sign of our times that experts disagree, fail to understand why they disagree, and become frustrated at the perpetuation rather than resolution of disputes over the effects of technology.

2.14 SUBOPTIMIZATION IN SYSTEM SAFETY ANALYSES

The accuracy of a quantified solution obtained by optimization depends upon the awareness of the various alternatives available and upon an understanding of their relationship to the system output, E_o. Solutions based upon incomplete criteria or an incomplete set of alternatives may be equivalent to correct answers to inappropriate questions. One subsystem may be optimized rather than the overall system by weighting one relevant factor too high in relation to another, equally important factor.

This is different from the situation in which one subsystem is intentionally weighted more strongly, a strategy which may be appropriate in a safety analysis. For example, it is not unusual to suboptimize about the human subsystem to assure safe operations in systems that require as combination of personnel and machine subsystems. While human factor considerations are intended to permit personnel to perform their tasks reliably and with ease, the ideal circumstance that permits activities to be performed within the system context is not always achieved. Rather, personnel functions are sometimes permitted to be carried out to the detriment of overall system performance. A presumption that this is done intentionally leads to the further presumption that risk and value judgments, rather than a mathematical optimization, are used to establish the weighting factor to be applied.

In practice, the process of optimization invariably reduces to one of suboptimization. To understand why this occurs, consider the implications of both processes. Complete optimization implies

1. Simultaneous consideration of all possible input alternatives and all possible outcomes, E_i's and E_o's.
2. Consideration of the effect all possible inputs, E_i's, have on the system.

The set of all possible E_i's includes those that are under the optimizer's control and all exogenous events, those not under the optimizer's control.

3. Optimization is subject to the initial constraints of the utility function selected by the optimizer. This could be the system safety engineer, the managers of the firm designing or producing the system, or the decision makers of some government agency. System safety optimization, therefore, may require a preliminary optimization of the input constraints.

Incomplete optimization, suboptimization, occurs when

1. Fewer than all possible alternatives are used to determine the effects upon E_o. It is usually not possible to consider the entire range of all possible alternatives, and it is often easy to overlook one crucial course of action which, if explicitly considered, might drastically change the allocations among the various alternatives.
2. It is usually not possible to deal with all the E_o's not under control of the optimizer. Consider, for example, the range of meteorological conditions that are used as design constraints in a passenger airplane to ensure a safe E_o. The totality of conditions is limited to fewer forces than those that can occur in nature and that are, indeed, sometimes encountered during operations.
3. Criteria that are used in actual problems have some degree of imperfection. Consequently, these criteria are approximate indicators that must fail to reflect the full range of possible E_o's.

Suboptimizations can be classified into three categories. First is the "one-dimensional" objective function in which there is either one important objective or, if there are several, all can be reduced to the equivalent of a single measure. Suppose, for example, that safe arrival is established as an objective in a passenger airline subsystem. In such a circumstance, all resources and all alternatives are directed toward ensuring that, regardless of the nature of E_i, the system's response is one that achieves the desired objective.

The second kind of suboptimization is multidimensional, but with the condition that all important system objectives may be quantified analytically. In the passenger airline system, the single objective of safe arrival could be made multidimensional by including additional objectives relating perhaps to maximizing return on investment, minimizing downtime and maintenance requirements, and optimizing flight plan scheduling to achieve the best mix of passengers and cargo.

The third type of optimization is multidimensional but contains certain objectives which, although possibly important, are omitted from the optimization process because they are not subject to quantitative analysis. In the passenger airline system, factors related to the selection of new cabin attendant uniforms, advertising, and selection of foods to be served fall into this category. While these three factors have no effect on safety, they may be of some importance in maximizing return on investment.

Ideally, the application of system safety methodology to the design and utilization of a system results in an optimum degree of safety. However, the assumption that suboptimization results in a safety level that is lower than the one obtainable by optimizing may be in error. There are occasions when an inability to obtain a quantitative, optimum value forces constraints on the system that result in an unnecessarily high safety level. For example, the Williamsburg Bridge, built across the East River in New York City in the late nineteenth century, was found in recent times to have a safety factor high enough to permit the addition of a second traffic level without severe degradation of the inherent safety.

2.15 THE DOMAIN OF SYSTEM SAFETY

Fulfilling the functions and responsibilities established for system safety requires a role that is far-reaching and interdisciplinary. It includes, for example, the prevention of injury or loss of property brought about by

1. Defects or inadequacies in design, material, or quality of work
2. Degradation caused by processes such as manufacturing, testing, and the stresses of transportation
3. Human errors of omission or commission in operating or maintaining a system
4. The environment acting upon the system at interfaces with other systems or with nature
5. The occurrence of credible accidents

Prevention of property loss and injury due to automobile accidents illustrates such an interdisciplinary aspect of system safety. One tends to think mainly in sympathetic terms about the victims of such accidents. It is interesting to consider, however, how the damage suit, used so often as a legal remedy in automobile accidents, has clogged the courts and imposed a $7 billion annual bill on liability insurance holders. Though this amount is large, it contributes little to highway safety. By redefining the system, $7 billion could perhaps provide more safety as well as still provide compensation for the victims of a diminished number of accidents. Considerable effort is being directed toward achieving this end, by such groups as the Insurance Institute for Highway Safety. No-fault insurance is felt by some also to be of merit for unclogging the courts.

2.15.1 Single Hazards

In general terms, the interdisciplinary character of system safety requires that all events capable of causing hazards be considered. These include mechanical and operational hazards, which are defined as follows:

A *mechanical hazard* is one which inhibits or prevents proper equipment operation through failure or degradation of one or more elements.

An *operational hazard* is one which does not cause equipment or system failure but which, nevertheless, inhibits or prevents the system from carrying out its intended function.

Solar radiation interference with radio communications is an operational hazard in that such radiation creates static that may block communications but does not damage the hardware.

An additional refinement in hazard classification can be made by distinguishing between endogenous and exogenous hazards:

An *endogenous hazard* is one caused by inherent defects in design, material, quality of work, or operating procedures.

An *exogenous hazard* is brought about by phenomena external to the system, such as lightning or cosmic radiation.

The fundamental distinction between these two is that endogenous failures are self-induced because they are internal to the system, and exogenous failures are externally induced. Either can cause a mechanical or an operational hazard.

2.15.2 Multiple Hazards

In addition to single hazards, it is necessary to consider the effects of various combinations of hazards in a safety analysis, in particular whether some or all of a certain group of hazards can combine to present a danger to the system when, on an individual basis, each does not.

Radiation hazards illustrate such a combination. Until three or so decades ago the average individual was exposed only to ambient levels of radiation in the environment (less than 1 millirem per hour) and to radiation required by medical and dental procedures. This condition was changed when the first atomic bomb was detonated; testing of atomic weapons has slightly increased the "ambient" background radiation. Efforts to prevent further changes in the natural radiation background in the atmosphere through control of aboveground atomic testing may not be 100 percent effective. In addition, there are some who claim that we do not fully understand the effects current underground atomic weapon testing will have several centuries from now.

Looking into the near future, humans can expect a more extensive interface with various sources of radiation, particularly with the increasing number of nuclear reactors used to generate electricity in many countries all over the earth. Availability of oil and natural gas may be severely limited for political and nationalistic reasons, and projections show that natural supplies of these fossil fuels may not be infinite. These concerns suggest that an increase in nuclear-generated electricity is a reasonable choice for the next few decades. However, safety problems that exist for one plant are multiplied as the number of nuclear power plants is increased. One such problem is core meltdown; that is, in the event a full-size nuclear reactor core suddenly loses coolant due to a failure in its primary cooling system, and the emergency systems also fail when called upon to operate, the most likely consequence would be melting of the nuclear reactor core. Subsequently, it can be expected that the reactor vessel and the outer, secondary containment walls of the plant would be breached and radiation would be released to the atmosphere. The interval of time between failure of the cooling system and release of radiation into the atmosphere would probably be less than one hour. A real-life drama associated with such an occurrence was initiated on March 28, 1979, as a result of a cooling system failure in the No. 2 unit of the General Public Utilities' Three Mile Island nuclear facility near Harrisburg, Pennsylvania.

Another problem associated with nuclear generation of electricity is the ever-increasing amount of radioactive waste generated by these energy sources, some of which has a half-life of 25,000 years. Recent discoveries in populated areas where radioactive waste was created by the production of radium during the early part of the twentieth century illustrate some of the consequences that can occur when all aspects of safety are not fully understood in this domain.

Another interface with radiation that needs to be considered now and for the near future is encountered at altitudes of 70,000 feet in supersonic transports, which are expected to proliferate within the next decade. At present, a person can be exposed to three millirems of radiation by taking a conventional round-trip airplane flight across the country. (For comparison, a single back-to-front X-ray imparts a dose of 10 to 30 millirems. The normal ambient, daily radiation dose from cosmic, terrestrial, and internal sources—food, water, and air—is about 0.342 millirems.) Studies made to establish optimum safety levels in supersonic transports may also be applicable to the hypersonic transports that are expected to be operational by the end of the twentieth century and, eventually, to commercial space travelers. With regard to the immediacy of the problem of excessive radiation in space for the average individual, it is noted without comment that Pan American World Airways stopped accepting reservations for excursion flights to the moon in early 1971, after 50,000 were received.

The first problem to be considered in a safety analysis of radiation hazards is the determination of a group of "acceptable" radiation levels. A group of levels is necessary because no single standard can serve. For example, in the case of a supersonic aircraft it might be appropriate to have two standards, one for the passengers and one for crew members, who, it can be presumed, will be exposed to the radiation hazard on a more routine basis. The notion of separate standards, however, requires a distinction between passengers who fly routinely and those who fly occasionally. On the other hand, one standard might place an unnecessary weight penalty on the aircraft for shielding.

It is possible, without undertaking a rigorous mathematical solution, to establish some boundaries for acceptable levels of radiation. The lower bound can be established as levels of radiation that are normally encountered in conventional, earthbound environments. An upper bound is more difficult to establish. A radiation level which is fatal is obviously too high. A level established by the risks some crew members and travelers might be willing to accept in supersonic and hypersonic flight might be higher than one society is willing to permit. Yet the nature of the vehicle's design and construction depends upon some specified level. How thick and of what material shall the hull be constructed? As the weight of the hull increases, the number of passengers and amount of cargo will have to decrease.

Consequently, the logic that dictates the value of an upper limit for radiation hazards needs to be considered both on a constituent basis and by combinations of two or more constituents. The method of analysis is illustrated with a hypothetical hypersonic traveler, who could be exposed to all possible types of radiation: electromagnetic radiation, particulate radiation, or both. Electromagnetic radiation includes gamma rays, photons, and X rays; particulate radiation includes alpha particles, electrons, heavy primary nuclei, neutrons, and protons. Secondary effects such as *bremsstrahlung* are not considered in this discussion.

There is a specific level of exposure for each of the electromagnetic and particulate forms of radiation that is fatal for any time interval of interest. Combinations of exposure at levels less than the fatal level established for each radiation form can also be fatal. Of concern in the combinatorial form is the amount of radiation, induced by two or more forms, that brings about an equivalent residual dose (erd) of 450 roentgen equivalent man (rem), for this amount of radiation is fatal 50 percent of the time. A further discussion of this subject and of a model which indicates how the various forms of radiation can be combined to yield 450 rem is presented in Chapters 6 to 8.

2.16 MANAGING TO ENSURE SYSTEM SAFETY

This chapter has related thus far to the nature of system safety and to the domain in which its activities are conducted. Actual results, achieving an optimum value or relatively acceptable safety level, depend to a large degree upon system safety management and system safety implementation, which are defined as follows:

> *System safety management* is that element of program management which ensures the accomplishment of the system safety tasks, including identification of system safety requirements; planning, organizing, and controlling those efforts which are directed toward achieving the safety goals; coordinating with other (system) program elements; and analyzing, reviewing, and evaluating the program to ensure effective and timely realization of the system safety objectives.

> *System safety implementation* consists of those activities carried out for the application of scientific principles needed for the timely identification of hazards and for the initiation of actions necessary to prevent or control hazards that are determined to be inherent in the system.

The responsibilities of management personnel and their ability to implement safety revolve about the tasks of specification, accomplishment, and measurement. The relationship of these tasks is shown in Figure 2-3. One quality of Figure 2-3 is that it is itself a model of a system with feedback. Consequently, the domain of system safety and the manner in which it relates to the system under design or in use is such that it can be treated as a system— in this instance, a safety management-implementation system. Some characteristics of this system are

Figure 2-3
Management-Implementation System

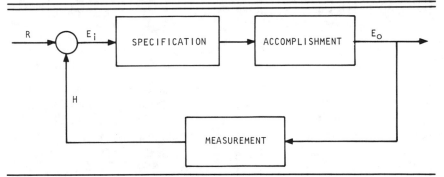

1. The objective of the management-implementation is to achieve an optimum degree of safety for the system being designed or used.
2. The system inputs, E_i, are represented by some combination of the feedback data, H, and the reference level, R.
3. The reference level is derived by value considerations of the desired output, E_o, which may be
   a. Specified, ideal, or optimum;
   b. The current or projected state of the art, which is itself a measure of the level of safety that can be achieved per dollar expended; or
   c. The value of the safety level already achieved by the system, H.

The management-implementation system is one of the forms that the generalized schematic representing a system may take. The inputs and outputs of the management-implementation system, E_i, and E_o, are generally functions of time. This type of system can be used as a model for conducting system safety studies by the designer, the manufacturer, and the user. If a user does not choose to maintain a separate management-implementation to ensure user safety, inputs can, nevertheless, be made known to the designer of the system for inclusion in the specification block. In some cases the user monitors the status of the implementation block employed by the designer and manufacturer and may make independent measurements of the safety level achieved or monitor the safety levels measured. Major airlines usually operate in this fashion and may even participate with the manufacturer in establishing safety criteria for the aircraft they purchase. Appliance manufacturers routinely test new and modified products in test markets for, among other things, safety considerations.

Feedback is necessary in the management-implementation system to permit the measurements of safety levels to be compared to a model, prepared in advance, of the growth of safety as a function of time and the status of the system life cycle. The notion of system life cycle is discussed in Section 4.2.1, and safety growth curves are considered in Section 2.17.

2.16.1 Management-Implementation System: Specification Block

The specification block in Figure 2-3 consists of the set of tasks performed to establish safety levels that are to be achieved by the system. In practice, this implies specification or allocation for safety-related functions for all of the elements of the system. Adjustments may be made to the specifications as measurements are obtained that indicate how well the specification requirements have been achieved. In the order in which they are generally conducted, the components of the specification block are the following:

1. System safety studies are made to establish a feasible, optimum, or desirable safety level for the system.
2. The safety level is adjusted by trade studies, which extend the effects of hazards on the system and personnel to their effects on equipment, cost, schedules, and other system parameters.
3. Safety studies are made at the subsystem and component levels in order to apportion the safety level established for the system to the subsystems and, in turn, to the components and elements.
4. Studies are made to determine the weighting to be given to hazards and injuries as a function of time. In a transportation system, for example, certain hazards occurring during earlier parts of a mission could prevent the system objectives from being accomplished. The identical hazards occurring near the completion of the mission might allow some of the system objectives to be achieved even though the effects of the hazards are serious.

Time enters in still another way on missions for which the possibility of successful abort or rescue is remote. "Points of no return" must, of course, be defined. One such point is interruption of oxygen flow to the brain for more than about four minutes, which need not cause death but can cause irreversible brain injury.

2.16.2 Accomplishment Block

The accomplishment block of Figure 2-3 consists of those tasks carried out for the purpose of implementing the tasks needed to achieve the requirements and specifications established for the system and its elements. For some systems these tasks are mainly qualitative, often requiring no more than adherence to sound, established industrial safety practices. This places an obligation on the safety engineer to generate a complete set of safety standards and practices and to provide the necessary training and motivation to all personnel to assure adherence. It is sometimes possible to adapt standards established for other, apparently dissimilar systems. For example, habitability studies conducted for atomic submarines and diving bell operations have provided data for establishing requirements for spacecraft environments.

In most instances, the accomplishment block is a mixture of qualitative and quantitative tasks. Quantitative tasks are undertaken whenever the magnitude of the hazard vector is large enough to warrant such concern. The needs must be determined separately for each system. In any event, the kernel of the effort contained in the accomplishment block consists of identifying hazards and completing the tasks necessary for their elimination or control.

2.16.3 Measurement Block

As its name suggests, the tasks in the measurement block are directed toward obtaining a measure of the safety level already achieved and of those predicted for the system. During early phases of system life cycle, the measure is an estimate obtained from available knowledge of the proposed system and augmented, where possible, by usage data obtained from similar systems. The confidence level of the initial estimate is increased, and the estimate is subsequently transformed into a measurement as design, test, and usage data are obtained for the various system elements. As Figure 2-3 implies, the results obtained from the measurement block, H, are compared to the requirement specified. A value of H equal to or greater than E_i implies one of two possibilities, namely:

1. The value of E_o was selected properly and the management-implementation system is achieving its objective. If H is much larger than E_o, then some of the resources allocated for achieving E_o can be redirected into other system efforts.
2. The management-implementation system is indeed operating so as to obtain the intended value of E_o, but this value was set too conservatively or at too low a level. The occurrence of this possibility is determined by comparing E_o to R.

If H is less than E_o, it is necessary to consider whether the specification was set at too high a level or whether additional resources need to be allocated to the management-implementation system.

2.17 GROWTH OF SYSTEM SAFETY

Learning for individuals—gaining knowledge and skill about the performance of some task—is a complex process depending upon many variables. Nevertheless, predictions can be made for the rate at which learning can be expected to occur. An analogy can be drawn between learning that occurs for an individual and the "learning" that occurs as the growth of the safety level of a system. The manner in which the safety level of a system can generally be expected to grow as a function of time is shown in Figure 2-4. The ordinate indicates the level of safety predicted or measured, ranging from 0 to 1.0. The abscissa relates to either the entire life of the system or to some portion of the life cycle. The dashed horizontal line, shown in the figure as an optimum, can be either a goal or a requirement.

The growth model presented in Figure 2-4 has the form of a Gompertz curve. That is,

Figure 2-4
Safety Growth Model

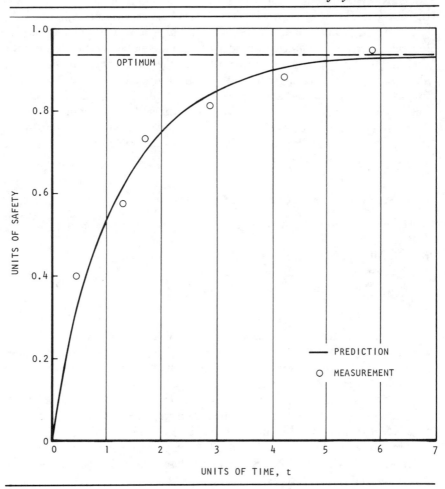

$$SL = ab^{c^t} \quad \text{for } 0<a<1; \quad 0<c<1 \tag{2.2}$$

where

SL = the safety level
a = the upper limit approach by safety as t increases to its limit
b = a constant
c = the slope of the curve
t = the period of time under consideration

This model was chosen because it has been shown to be applicable in other domains where growth and learning follow a similar process. The changes in the safety level that occur during the system life cycle are brought about by either modifications made to the system, changes in design, material, or inspection procedures, for example. The graph is based on the assumption that major changes are not attempted in the time span shown on the abscissa. Such attempts would introduce a discontinuity in the curve, lowering the value of the safety level at the time the changes are introduced. Safety growth models may also show "improvement" as a result of refinement of the measurement process. This occurs because initial measurements made, perhaps, when not all data are available tend to be treated in a conservative fashion by the safety analyst.

2.18 NATURE OF SAFETY DOMAIN

It has been shown thus far that system safety is concerned with

1.   Identification of hazards
2.   Determination of optimum or acceptable safety levels
3.   Elimination or minimization of known hazards

These three responsibilities complement one another. In the ideal circumstance, all possible hazards concerning a given venture, along with their likelihood of occurrence, are known a priori. The safety level is then known with certainty and all that remains is to determine whether the risk is acceptable, regardless of whether the level is optimum. In the event that the safety level is assessed to be too low, it may be presumed that an increased allocation of system safety management and implementation resources will be applied in an attempt to increase the level of safety achieved. However, system safety management and its implementation do not operate independently of other system activities. In fact, these interfaces always affect the decisions and optimizations attempted by safety. A discussion of these interfaces and their relationship to system safety is presented in the next chapter.

In addition to growth in the safety domain based upon new data and technology (vertical growth) or perhaps as a consequence of such growth, safety is also expanding horizontally. Areas of specialization in this field include

1.   Industrial safety, concerned primarily with environmental health, occupational hazards, and fire protection in the workplace.
2.   Product (consumer) safety, directed toward safe operation and disposal

of products and substances and ensuring users against unsafe design qualities.

3. Mission safety, military or civilian, oriented toward accomplishing a task with minimal risk to personnel and equipment. Initially a military concern, this technology is now applicable to many scientific endeavors, search-and-rescue activities, and transportation systems, to name a few.

4. Personal safety, concerned with safeguards provided for and practiced by individuals. A recent example is the formation of Health Maintenance Organizations (HMOs) whose profits improve if illness is prevented among their members.

The technology of system safety is gaining recognition as a useful means of analysis for these and other safety specialties.

Additional Reading

Christianson, D. Risk comparison. *IEEE Spectrum,* September 1979, p. 25.

Ellis, D. O., and F. J. Ludwig. *Systems Philosophy.* Space Technology Series. Englewood Cliffs, N.J.: Prentice-Hall, 1962.

Foster, J. W. Systems analysis as a tool for urban planning. *IEEE Spectrum,* January 1971, p. 48–54.

Kampus, K. C., and L. R. Lamberson. *Reliability in Engineering Design.* New York: Wiley, 1977.

Lederer, J. The modes of safety. *Hazard Prevention,* November–December 1979, p. 8–9, 32.

Llacer, J. M. Nuclear Medical Forging, *IEEE Spectrum*, July 1981, pp. 33–37.

Lynn, P. Danger: Leaks in gas-pipe safety. *Wall Street Journal,* June 3, 1971, p. 16.

McCloskey, J. F. and F. N. Trefetethen. *Operations Research for Management.* Baltimore: Johns Hopkins Press, 1956.

Maynard, H. B. *Industrial Engineering Handbook.* New York: McGraw-Hill, 1977.

Miles, L. D. *Techniques of Value Analysis and Engineering.* New York: McGraw-Hill, 1961.

Newman, J. R. *The World of Mathematics.* New York: Simon & Schuster, 1956.

Productivity. *IEEE Spectrum* (special issue), October 1978.

Smith, C. O. *Introduction to Reliability in Design.* New York: McGraw-Hill, 1976.

Wildavsky, A. No risk is the highest risk of all. *American Scientist,* January–February 1979, p. 32–37.

Chapter 3

SYSTEM SAFETY INTERFACES–OPERATIONS WITH EVENTS

3.1 SYSTEM EFFECTIVENESS

As indicated in previous sections, system safety interfaces with all disciplines that are involved in the design, development, and operation of the system under consideration. Before discussing these interfaces from the safety point of view, it is desirable to describe the relationship between system safety and the system from the system point of view. The mechanism most used for this purpose, one that is considered to provide a good suboptimization for system objectives, is often referred to as *effectiveness, mission effectiveness,* or *system effectiveness.* In the main these terms are used in the same context. The one chosen for use in this text is system effectiveness, *E*, which is defined as follows:

> *Effectiveness* is the measure of the extent to which a system may be expected to achieve a set of stated system objectives.

The disciplines usually considered in the suboptimization of *E* and their abbreviations are shown in Table 3-1. Early efforts in relating these disciplines to system effectiveness was carried out by the Weapon System Effectiveness Industry Advisory Committee, a group established in 1964 to provide technical guidance and assistance to the Air Force Systems Command. This advisory group was assigned five objectives, one of which was to "develop a basic set of instructions and procedures for conducting an analysis for System Optimization considering effectiveness, schedules, and funding."

In general form the functional relationship between $E(t)$ and the six disciplines listed in Table 3-1 can be written

$$E(t) = f\left[\left(\frac{S_a}{S_s}\right), \left(\frac{R_a}{R_s}\right), \left(\frac{M_a}{M_s}\right), \left(\frac{Q_a}{Q_s}\right), \left(\frac{H_a}{H_s}\right), \left(\frac{V_a}{V_s}\right)\right], \qquad (3.1)$$

Table 3-1
Components of Effectiveness

Discipline	Abbreviation
Safety	S
Reliability	R
Maintainability	M
Quality assurance	Q
Human factors	H
Value engineering	V

where

    a = the achieved level of each parameter at some specified time in the system's life

    s = the specified level established for that parameter

The functional relationship expressed by Equation 3.1 needs to be written as an explicit expression if a value for E is to be obtained at some point in time. No one explicit expression can be proposed, however, for $E(t)$ depends upon factors that are unique to each system. Values specified in the denominators of the factors of E in Equation 3.1 can relate to one steady state quantity, such as that intended for use as an objective at the end of a program, or to a variable objective, such as that defined during early steps of a program by a growth curve. The value achieved is given by the numerators of the various components of E. Quantification of achieved values is obtained from data available through testing or usage, by measurement, or by estimation.

One problem that increases the complexity of explicitly expressing E is that the components of E are almost never completely independent of each other. In addition, it may be necessary to substitute other parameters related to E in lieu of the set of six presented in Table 3-1. For example, availability is sometimes used in place of maintainability, and survivability is occasionally used in lieu of reliability or safety. Another problem in explicitly expressing E is that its components have different utility values, k_i. When these are known or can be estimated, Equation 3.1 can be written

$$E(t) = f\left[\left(\frac{S_a}{S_s}\right)^{k_1}, \ \left(\frac{R_a}{R_s}\right)^{k_2}, \ \left(\frac{M_a}{M_s}\right)^{k_3}, \ \ldots \right] \text{ for } 0 \leqslant k_i \leqslant 1 \qquad (3.2)$$

Table 3-2 shows some of the resources and variables that may be used for trade-offs in attempting to optimize the system effectiveness of a

Table 3-2
Typical Trade Elements in a Communications Satellite

Resources	Variables	Alternatives
Funding	Safety	Manned vs. unmanned
Time	Maintainability	Synchronous orbit vs. random
Facilities	Quality assurance	Wide band for few channels vs. narrow band for many
Personnel	Human factors	Battery vs. fuel cells vs. solar cells vs. atomic source
Special tooling	Reliability	Redundancy vs. maintainability vs. high-reliability parts
Payload weight	Value engineering	Method of station keeping, types of stabilization

hypothetical communications satellite. Consider for example the first alternative, manned vs. unmanned. A manned satellite places greater requirements on training and training facilities and, possibly, on initial time and funding. All other factors being kept equal, safety requirements will be increased but maintainability will be enhanced. As a consequence of superior maintainability, it is possible that value may be enhanced. A more detailed discussion of the interfaces between system safety and the remaining disciplines listed in Table 3-1 is presented in the next few sections.

3.2 INTERFACE WITH RELIABILITY

System safety is more closely related to reliability than to any of the other disciplines defined by *E*. The basis for this interface can be seen by examining the definitions of the two disciplines. The generally accepted definition of *reliability*, applicable to both hardware and human systems, is

> The probability that a system will perform its intended function for a specified period of time under a set of specified environmental conditions.

Time and environmental conditions must be specified if quantification is to have precise meaning. The definition of *safety*, presented in Section 2.1, is

> Freedom from conditions that can cause injury, illness, or death to personnel, damage to or loss of equipment or property, or environmental harm.

Let us disregard for the moment that this definition for safety is qualitative rather than probabilistic. It can then be seen that hazards which do not cause injury or death can fall into either the safety or the reliability domain. Further, it is reasonable to assume that injury or fatality can result from inability of a system to perform its intended function, a reliability concern. Conversely, the occurrence of a hazard which affects only personnel, a safety concern, can as a secondary effect be responsible for preventing a system from performing its intended function, thereby degrading the reliability of the system.

  This closeness between safety and reliability, plus some analogous relationships with other allied disciplines, can tend to diffuse the interface between safety and reliability. Three other qualities which inhibit a precise definition of this interface are the following:

1. Both safety and reliability are dynamic disciplines. That is, improvements are continuously being made in existing techniques and procedures, new methodologies are continuously being created, and both established and new technologies are constantly being applied to new areas.
2. The general-purpose management-implementation system for safety and reliability introduced in Section 2.16, which relates specification, accomplishment, and measurement and transmits data to all three by feedback, represents the general case. This general case needs to be modified for each system under analysis to ensure a unique, one-to-one relationship between the actual functional systems and the blocks of the general-purpose management-implementation system.
3. Relationships established for reliability and safety are functions of time. Consequently, the mathematical description of such relationships can differ from instant to instant.

  Nevertheless, it is possible to define some characteristics of the safety-reliability interface. As a prerequisite, it is necessary to arrange for both to be directly comparable by expressing them in the same units. This is achieved by transforming safety into probabilistic terms; that is, safety is defined in terms of the probability that injury or damage does not occur. Once this transformation is achieved, Boolean notation and Venn diagrams may be used to describe both disciplines. A discussion of Boolean notation is presented next; Venn diagrams are discussed in Section 3.2.5.

3.2.1. Sets

The first notion to be considered in a discussion of Boolean algebra is that of a set or class. The term *set* is used to denote any well-defined collection of entities. The collection may consist of

1. Real entities, whose occurrence has happened or is known to be a certainty in the future
2. Potential entities, whose occurrence is desirable and, consequently, sought after
3. Undesirable entities, whose occurrence is to be prevented

The set of all possible outcomes that can be obtained by rolling a die is 1, 2, 3, 4, 5, 6. The set of all possible outcomes in the domain of safety obtained by leaving home in the morning and driving to work is safe arrival and no damage to car, damage to car but no injury, injury but no damage to car, injury and damage to car.

Each member of a set is known as a *sample point*, and the collection of all possible sample points is known as a *sample space* or the *universal set*. The set of sample points in a sample space can be infinite, such as all the integers or all the points on a line, or it may be finite (discrete), as in the die and car examples.

3.2.2. Events

Consider now the notion of an "event" which, like a set, is also defined as a collection of well-defined sample points. The distinction between set and event is that sets are so general in nature and are so prevalent that they may be considered to afford the logical foundation for all mathematics. The notion of a sample space derives its importance from the fact that it provides a means to define events. The mathematical formulation of an event depends upon the fact that, for any set of events under consideration, there is a set of descriptions such that an event can occur if and only if the observed outcome has a description that lies in the set. For example, the event "7," one of the possible outcomes obtainable by throwing a pair of dice, is defined by describing the set of sample points in the universal set whose description gives the desired result.

Consequently, an event may be defined as a set of descriptions, such that if it is claimed that an event *E* has occurred, it implies that the outcome of the situation under consideration has a description that is also a member of *E*. Two notions are involved, both central to safety considerations; namely, the notion of "event" and the notion of "occurrence of an event." Both are needed in safety analyses because hazards are defined (Section 2.3) as potential conditions that may result in damage or injury. The concept of "event" is used in safety studies for the construction of mathematical models. The notion of "occurrence of an event" is used for translating statements made in the mathematical model into statements about the real world. The purpose of such modeling is prevention or control of those events considered undesirable.

3.2.3. Simple and Compound Events

Events may be classified as simple or compound. A compound event is one that can be subdivided into two or more simple events; a simple event cannot be divided any further. For example, the occurrence of a 1 in the throw of a die is a simple event. The occurrence of a 7 in the throw of two dice is a compound event, for this can be divided into two simple events, equivalent to throwing one die twice. Each set of descriptions—(6, 1), (5, 2), (4, 3), (3, 4), (2, 5), and (1, 6)—satisfies the event "occurrence of 7."

A simple event may be represented by one sample point in a sample space. The sample space defined by the sample points represented by the faces on a die consists of the six points, 1, 2, 3, 4, 5, 6 on an axis of real numbers. The sample space I, representing the possible outcomes obtainable by throwing two dice, is shown by Figure 3-1, a square 36-point lattice. In this figure, the x and y axes represent two identical sample spaces, each obtained by listing the sample points represented by the faces on one die. Each point of the lattice represents one of the possible simple events obtainable by one throw of the dice. The events enclosed in Figure 3-1 define the descriptions of the event 7.

A sample space containing an infinite number of sample points is illustrated by Figure 3-2. It represents the following example. A casualty underwriter is interested in the accumulated time x_1 and x_2, respectively, on two critical components used to ensure the safety of a large system. Suppose all critical components are aged for six months before installation in the system and that each is replaced after six years. The sample space for the event "usage times for x_1 and x_2 at any time after installation and prior to replacement" is defined by the area $abcd$ in Figure 3-2. The line ac represents the event "usage times of both components are equal," and the compound event "component x_1 has more usage time than x_2" is defined by the ruled area below line ac.

The type of diagram represented by Figures 3-1 and 3-2 is modified somewhat and utilized in Section 3.2.5 to describe the interface between safety and reliability.

3.2.4. The Safe Event

The event that no injury occurs to personnel or that no damage occurs to property over some specified period of time in the system life cycle is of primary concern in the domain of safety. For a complex system, a "womb to tomb" life cycle can be divided into five facets: concept formulation, contract definition, development, production, and operations. The relationship of system safety to each of these facets is developed in Chapter 4. The universe

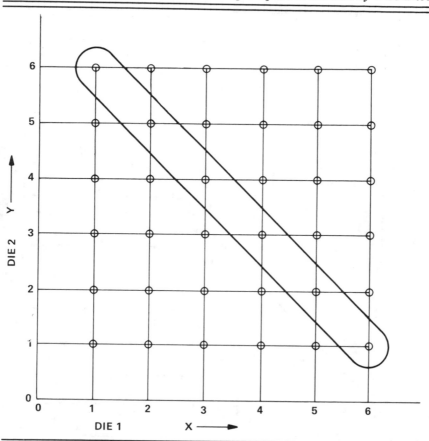

Figure 3-1
Sample Space Described by Two Dice

in which the safe event is formally defined can be established over the entire life cycle of the system or it may be restricted to some portion of the life cycle, such as production or operations.

As an example, the universal set in the example of driving to work (Section 3.2.1) contains four sample points. The collection making up the safe event contains one sample point, and the one for the unsafe event contains three sample points. Having defined the safe event S, it is natural to be concerned with the probability that S will occur, written $P(S)$. To do this it needs to be known that for any event E there is an event $\bar{E}$, called the complement of E, which is the event that E does not occur. $\bar{E}$ consists of all the descriptions in the universal set that are not in E. Of concern in safety are

Figure 3-2
Compound Events in a Continuous Sample Space

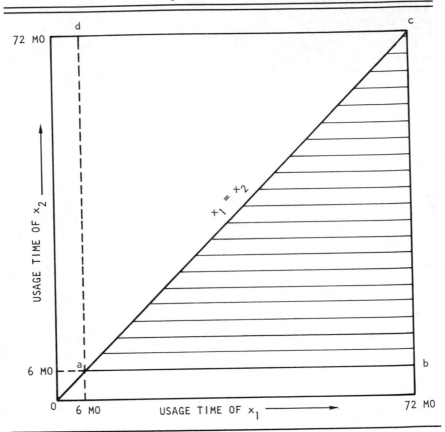

the probabilities that S will not occur or that $\bar{S}$ will occur, which is written $P(\bar{S})$. Inasmuch as a universal sample space by definition contains all possible outcomes, the sum of the probabilities that either the safe or unsafe event will occur is unity. That is,

$$P(S) + P(\bar{S}) = 1 \qquad (3.3)$$

With Equation 3.3 as a basis, a definition of safety may now be given which corresponds to the probabilistic definition of reliability presented in the first paragraph of Section 3.2. Namely, *safety* is equal to the probability that the event S will occur, where the probabilities of the events S, $\bar{S}$, $P(S)$, and $P(\bar{S})$ are related as shown in Equation 3.3.

3.2.5 The Venn Diagram

It is possible to represent two sets dealing, for example, with safety and reliability in a diagrammatical fashion so that their interfaces can be readily visualized, as in Figure 3-3. The idea of using diagrams like this one seems to have originated with the Swiss mathematician Leonhard Euler (1707–1783). Subsequently, the British logician John Venn (1834–1923) made a number of refinements, and such diagrams are now referred to as Venn diagrams.

The universal set for the safety-reliability events of concern in a system may be represented by the rectangle of Figure 3-3. It contains a finite number of sample points, those which define the safe event S, the unsafe event $\overline{S}$, the reliable event R, and the unreliable event $\overline{R}$. Each of these four events consists of a defined collection of sample points, and each is a

Figure 3-3
Safety-Reliability Interface

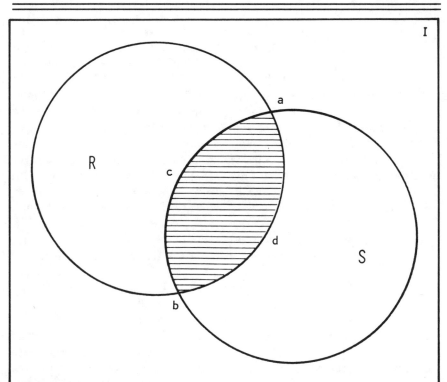

subset that is wholly contained in the universe I. "S is contained in I" is written $S \subset I$. Geometrically, $S \subset I$ and $R \subset I$ are represented by the two circles in Figure 3.3. The notion that a sample point x_i is an element of S is written $x_i \epsilon S$. The interface between safety and reliability in Figure 3-3 is represented by the lined area bounded between the arc acb, the extension of the safety event into the reliability event, and the arc adb, the extension of the reliability event into the safety event. There are two implications apparent from an examination of Figure 3-3:

1. $\bar{R}$, the unreliable event, represented by the area outside the R event, includes sample points that are in the safe event. In the driving to work example in Section 3.2 such a sample point may be illustrated by a carburetor malfunction that prevents the automobile from being operational but does not cause an accident.
2. $\bar{S}$, the unsafe event, represented by the area outside S, includes sample points that are contained in R. In the same example this may be illustrated by a sample point representing a collision caused by careless driving.

The sample points common to S and R define a subset of I called the "intersection of S and R," which is denoted by $S \cap R$. Formally, the event $S \cap R$ can occur if and only if both S and R occur, such that the observed outcome has a description that is a member of both S and R. Therefore, if $x \epsilon S$ and $x \epsilon R$, then $x \epsilon S \cap R$, and, conversely, if $x \epsilon S \cap R$, then $x \epsilon S$ and $x \epsilon R$.

It might be presumed from an examination of the figure that the common goal of both safety and reliability is to expand the area defined by $S \cap R$ until $S \cap R = I$. The tasks required of system safety to implement this goal are discussed in the next chapter. It should be noted that adding other events to I does not imply that any of them will have an intersection with events already in I. If events A and B have no sample points in common, they are "disjoint." This is written $A \cap B = \Phi$, where Φ is referred to as the "null set."

It can be concluded from this discussion of the safety-reliability interface that, in general, any improvement made in the reliability of a system will result in a corresponding improvement in the safety of the system. Consequently, it is necessary for safety analysts to be knowledgeable of those aspects of reliability that relate most directly to safety enhancement. The most fundamental of such aspects concerns failure or degradation of system elements that can result in a hazard to the system or personnel. In studying this aspect of the safety-reliability interface, it is necessary to consider both element failure (or degradation) and the frequency of occurrence of such failure.

3.2.6 Survivability

Survivability, a variant of reliability, is often applied as a system measure in lieu of reliability. *Survivability* may be defined as

> The measure of the degree to which a system will withstand the environment in which it is placed and not suffer abortive impairment of its ability to accomplish the designated mission.

The fundamental distinction between reliability and survivability is that the former relates to activity carried out prior to the appearance of failure or degradation in accordance with established standards, while the latter relates to system activities conducted either before or subsequent to the occurrence of failure or degradation. An element of a system or, for that matter, an entire system, can be unreliable and still survive. The application of the notion of survivability in relation to the human system is obvious. The intent is to assure survivability, preferably with no degradation of system capability, despite the loss of a redundant element or the degradation of nonredundant elements.

Efforts carried out to assure survivability of a human system in the event of failure of one of its elements may or may not be dependent upon efforts carried out to assure the reliability of those elements. The selection of reliability, survivability, or both for establishing a set of component or system criteria necessary to assure safety depends upon individual circumstances. Consider, for example, a set of design and safety constraints placed upon a multi-engine aircraft. On the one hand, reliability requirements need to be established for each engine to ensure the safety of the system. In turn, these requirements affect the design and maintenance policies established for the engines. On the other hand, survivability criteria could logically be established for the system in the event of failure of one engine. These criteria also affect the design of the engines and possibly their maintenance. Where conflict arises between the two, it is necessary for compromise to be reached through optimization.

3.3 OPERATIONS WITH EVENTS

It has been stated that the study of system safety and its interfaces with other disciplines requires analysis of events and their occurrence. One procedure for learning about events is to design a set of experiments to be conducted at appropriate times during the system life cycle, to ensure that knowledge needed about the system will be available when required. Since knowledge cannot be obtained with 100 percent certainty about events and their likelihood of occurrence, it is necessary to accept a design that leaves some residual risk in the operations of a system. Consequently, safety considerations must place some reliance on statistics and probability in order to assess the amount of uncertainty and, therefore, the amount of residual risk remaining to the system and its operators or users. These probabilistic considerations are discussed in Chapter 9. The remainder of this section is concerned with some of the techniques used to operate with events.

3.3.1 Type of Operations

Consider a set of sample spaces or events $A, B, C, \ldots$. Let A consist of sample points $a_1, a_2, \ldots, a_n$; B of sample points $b_1, b_2, \ldots, b_n$; $\ldots$ or change to $a_n, \ldots$. Sample points, events, and sample spaces are manipulated by means of the two binary operations $\cup$ (union) and $\cap$ (intersection) and by the unary operation — (complement).

Manipulation by means of the operation intersection was defined in Section 3.2.5. The sample space or event resulting from the union of sample spaces or events A and B, written $A \cup B$, occurs when either A or B occurs. That is, the observed outcome of $A \cup B$ has a description that is a member of A or B, or both. Therefore, if $x \epsilon A$ or $x \epsilon B$, $x \epsilon A \cup B$ and, conversely, if $x \epsilon A \cup B$, then $x \epsilon A$ or $x \epsilon B$. Given A and B, the operations $\cup$ and $\cap$ describe whether both A and B, or at least one, and possibly both have occurred. The unary operation complement is governed by two laws:

$$A \cup \overline{A} = \omega \qquad (3.4a)$$
$$A \cap \overline{A} = \Phi \qquad (3.4b)$$

In these equations, ω (also referred to as I in the literature) contains all possible sample points and is known as the universal set. Φ is known as the null set or impossible event and contains no sample points.

Venn diagrams for some of the operations not represented by Figure 3-3 are presented in Figure 3-4.

1. The ruled area in Figure 3-4a represents the sample points in both A and $\overline{A}$.
2. In Figure 3-4b, the event A is contained in the event B, which is contained in I.
3. The ruled areas in Figure 3-4c and 3-4d represent the set of sample points that are contained in events A or B. In (c) event A intersects B, and in (d) A and B are disjoint.

To illustrate the notions of complement, union, and intersection numerically, let ω be all possible outcomes of the throwing of a die, i.e., $\omega = (1, 2, 3, 4, 5, 6)$. Now, let $A = (1, 2)$; $B = (3, 4, 5, 6)$ and $C = (1, 3, 5)$. It can then be determined, that:

$$\bar{A} = (3, 4, 5, 6)$$
$$\bar{B} = (1, 2)$$
$$\bar{C} = (2, 4, 6)$$
$$A \cup B = (1, 2, 3, 4, 5, 6)$$
$$A \cap B = \Phi$$
$$\bar{B} \cap \bar{C} = (2)$$
$$C \cup \bar{C} = (1, 2, 3, 4, 5, 6)$$
$$C \cap \bar{C} = \Phi$$
$$A \cup (B \cap C) = (1, 2, 3, 5)$$

Figure 3-4
Venn Diagrams

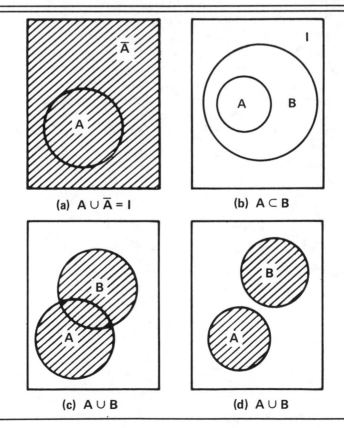

(a) $A \cup \bar{A} = I$

(b) $A \subset B$

(c) $A \cup B$

(d) $A \cup B$

3.3.2 Postulates Defining Operations with Events

Each branch of mathematics is based upon a set of undefined terms and unproven postulates. For example, the finite projective geometry developed by Veblen and Bussey uses two undefined terms, *point* and *line*, and five unproven postulates. The first of these postulates is

> The set contains a finite number of points. It contains one or more subsets called lines, each of which contains at least three points.

Events may be dealt with conveniently by Boolean algebra, which was developed by the English mathematician George Boole (1815–1864). However, before presenting some postulates by which sample points and events are manipulated in Boolean algebra, we will consider the principle of duality, which allows new statements and theorems to be obtained without formal proof. This occurs when one or more undefined terms can be substituted for other undefined terms without modification of the fundamental postulates in a branch of mathematics. In the finite projective geometry mentioned above, the principle of duality permits conclusions drawn about points to be re-written with the substitution of *line* for *point*. For example, it is true in projective geometry (as in Euclidean geometry) that two points determine a line. It is also true in projective geometry that any two lines determine a point. No formal proof is necessary of this last statement for it is obtainable from the previous statement by substituting *line* for *point* and *point* for *line*. The existence of the principle of duality in a branch of mathematics permits each statement to be derived from its dual.

In Boolean algebra, sample points and events are manipulated in accordance with four postulates. The dual of each postulate is also a postulate, so they are listed in pairs.

1. Distributive laws
$$A \cup (B \cap C) = (A \cup B) \cap (A \cup C)$$
$$A \cap (B \cup C) = (A \cap B) \cup (A \cap C) \tag{3.5}$$

2. Commutative laws
$$A \cup B = B \cup A$$
$$A \cap B = B \cap A \tag{3.6}$$

3. Associative laws
$$(A \cup B) \cup C = A \cup (B \cup C)$$
$$(A \cap B) \cap C = A \cap (B \cap C) \tag{3.7}$$

4. Idempotency laws

$$A \cup A = A$$

$$A \cap A = A \tag{3.8}$$

*In addition, there are two identity elements, I (or ω) and Φ which behave
in relation to union and intersection as follows:*

$$A \cup I = I$$

$$A \cap I = A \tag{3.9}$$

and

$$A \cup \Phi = A$$

$$A \cap \Phi = \Phi \tag{3.9a}$$

These four postulates and dentity relationships may also be expressed by
Venn diagrams; the diagrams for the distributive laws are shown in Figure
3.5.

3.3.3 Some Theorems for Operating with Events

Following are some theorems of general usefulness for operating with events,
all of them derivable from the postulates and identity relations.

$$(\bar{\bar{A}}) = A \tag{3.10}$$

$$\bar{\Phi} = I \tag{3.11a}$$

$$\bar{I} = \Phi \tag{3.11b}$$

$$A \cup (\bar{A} \cap B) = A \cup B \tag{3.12}$$

$$\overline{(A \cap B)} = \bar{A} \cup \bar{B} \tag{3.13a}$$

$$\overline{(A \cup B)} = \bar{A} \cap \bar{B} \tag{3.13b}$$

$$A \cup B = \Phi \quad \text{Implies } A = B = \Phi \tag{3.14a}$$

$$A \cap B = I \quad \text{Implies } A = B = I \tag{3.14b}$$

$$(A \cap B) \cup (B \cap C) \cup (C \cap A) = (A \cap B) \cup (C \cap \bar{A}) \tag{3.15a}$$

$$(A \cup B) \cap (B \cup C) \cap (C \cup \bar{A}) = (A \cup B) \cap (C \cup \bar{A}) \tag{3.15b}$$

$$\overline{(A \cap B) \cup (A \cap \bar{B})} = (A \cap B) \cup (\bar{A} \cap \bar{B}) \tag{3.16}$$

It is recommended that the reader derive some of these expressions and
express them in Venn diagram form.

Figure 3-5
Distributive Laws

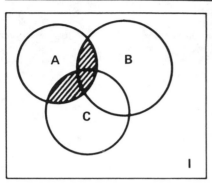

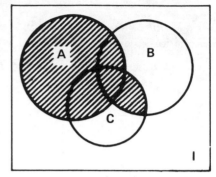

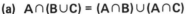

(a) A∩(B∪C) = (A∩B)∪(A∩C)   **(b) A∪(B∩C) = (A∪B)∩(A∪C)**

3.3.3.1 Absorption Laws

The dual equations

$$A \cup (A \cap B) = A \qquad\qquad (3.17a)$$

$$A \cap (A \cup B) = A \qquad\qquad (3.17b)$$

are known as the laws of absorption. Equation 3.17a may be proved as follows:

$$A \cup (A \cap B) = (A \cap I) \cup (A \cap B) \qquad \text{Identity 3.9}$$
$$= A \cap (I \cup B) \qquad \text{Distributive law}$$
$$= A \cap (B \cup I) \qquad \text{Commutative law}$$
$$= A \cap I \qquad \text{Identity 3.9}$$
$$= A \qquad \text{Identity 3.9}$$

3.3.3.2 de Morgan's Laws

A property of event operations frequently used in the safety domain, a generalization of the dual Equations 3.13a and 3.13b, is known as de Morgan's laws. The general form may be written

$$\overline{(A_1 \cap A_2 \cap \ldots \cap A_n)} = \bar{A}_1 \cup \bar{A}_2 \cup \ldots \cup \bar{A}_n \qquad (3.18a)$$

$$\overline{(A_1 \cup A_2 \cup \ldots \cup A_n)} = \bar{A}_1 \cap \bar{A}_2 \ldots \cap \bar{A}_n \qquad (3.18b)$$

Intuitive proof of de Morgan's laws can be obtained by operating with sample points. For example, consider $(\overline{A \cup B}) = \overline{A} \cap \overline{B}$. Suppose there are sample points $a_1, a_2 \ldots, a_i \in A$; $b_1, b_2, \ldots, b_i \in B$; and $c_1, c_2, \ldots c_i \in C$, such that $A \cup B \cup C = \omega$. Now consider $(A \cup B)$.

1. $A \cup B$ consists of the sample points $a_1, \ldots, a_i, b_1, \ldots, b_i$.
2. $(\overline{A \cup B})$, therefore, consists of the other sample points, $c_1, \ldots, c_i$.

Consider now, $\overline{A} \cap \overline{B}$:

1. $\overline{A}$ consists of the sample points $b_1, \ldots, b_i, c_1, \ldots, c_i$.
2. $\overline{B}$ consists of the sample points $a_1, \ldots, a_i, c_1, \ldots, c_i$.
3. $\overline{A} \cap \overline{B}$ consists, therefore, of the sample points $c_1, \ldots, c_i$.

Therefore, $(\overline{A \cup B}) = \overline{A} \cap \overline{B}$

3.4 INTERFACE WITH MAINTAINABILITY

Maintainability is a characteristic of system design, installation, and operations which may be defined for both hardware and human systems as follows:

> *Maintainability* is the probability that a system will be retained in, or restored to, a specific condition within a given period of time, presuming that required maintenance is performed in accordance with a set of prescribed procedures and allocated resources.

The term maintenance used in this definition may be defined as follows:

> *Maintenance* includes all actions necessary for retaining a system in, or restoring it to, a specified condition.

Maintenance conducted for the purpose of retaining a system in sound operating condition is preventive in nature. In general, this is carried out in accordance with some established schedule. Maintenance conducted for the purpose of restoring a system to an operational state implies that some difficulty has arisen and is considered to be corrective in nature. When carrying out preventive maintenance, it is not uncommon to uncover defects which require that corrective maintenance procedures be applied.

Since the definition of maintainability is expressed as a probability, its interfaces with safety and reliability can be described by a Venn diagram, such as the one presented in Figure 3.6. In this, all the relationships between S, R, and their complements are the same as in Figure 3.3. The interface between M and S is represented in Figure 3-6 by the arc *cdf*, and the interface between M and R is represented by the arc *ecd*. The area common to all three events,

Figure 3-6
Safety-Reliability-Maintainability Interface

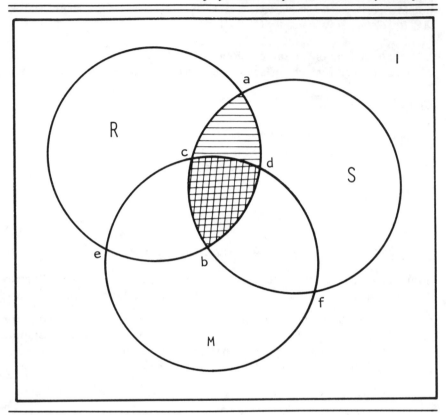

$S \cap R \cap M$, is represented by the cross-hatched area bounded by the arcs *bc*, *cd* and *db*.

Several relationships can be observed by examination of Figure 3-6. One, for example, is that not all the sample points in the subset $M \cap R$ are in *S*. This is due to the fact that the fundamental role of maintainability is to increase system life without necessarily enhancing safety. As a consequence, the utility of maintainability to a system is enhanced as

1. It becomes more expensive to replace the system, rather than to keep it maintained.
2. Achieving longer system life through improved reliability or redundancy of parts becomes less cost effective than conducting maintenance activities.

In some systems maintenance is closely allied with resupply. In the case of systems that are remote from supplies, such as spacecraft and surface and underwater ships, attaining long operational life, for example, may be dependent upon the ability to resupply certain expendables, such as propellants and food. A similar circumstance exists for humans in that system life can be degraded, or terminated, if something interferes with any one of the supply of expendables required to sustain life.

As a kind of converse, some sample points in the area defined by $S \cap M$ presume that maintenance is possible. Others, in $S \cap R$, presume that the sample points defining contingency activities, such as escape or rescue, are reliable. The following guidelines are offered to assist in determining whether a given sample point is contained in $S \cap M$, $S \cap R$, or $S \cap R \cap M$.

1. Direct removal and replacement of faulty components, or their repair by personnel *in situ*, is contained in $S \cap M$.
2. Switching to redundant equipment through remote means such as telemetry or *in situ* by attending personnel is contained in $S \cap M$.
3. Switching to redundant equipment through the use of built-in, self-checking circuits is contained in $S \cap M \cap R$.
4. Redundancy used in majority voting or in a fail-safe configuration for replicated elements is contained in $S \cap R$.

The process of idealizing the interrelationship described by Figure 3-3 suggests an expansion of the R and S sample points in I so that $S \cap R = I$. Although there are sample points located both in M and in R in Figure 3-6 that permit the event S to occur, this process of idealizing can nevertheless be extended to $R \cap S \cap M$. The ability to achieve this ideal, however, becomes more difficult as more parameters are added to the common intersection. An analogy to the ideal $S \cap R = I$ can be established for variables S, R, and M by permitting the union of either intersection with safety to fill the universe. That is, the ideal circumstance for R, M, and S may be expressed

$$(S \cap R) \cup (S \cap M) = I \tag{3.19}$$

The left-hand side of Equation 3.19 is represented in Figure 3-6 by the area bounded by the arcs *acbf*, *fd*, and *da*.

3.5 AVAILABILITY

Sometimes, the notion of availability, which relates to both reliability and maintainability, is of greater concern to safety than either reliability or maintainability, although it relates to those two concepts. Availability A, may be defined as follows:

Availability is the probability that a given system is in an operable state and can be committed at a given instant of time.

The interface between availability and safety is analogous to the one between reliability and safety and can be identified by listing the sample points that are outside both the availability and the safety universes. The relationship of the availability universe to both reliability and maintainability is directly derivable from the definitions for these terms. It can be expressed as follows:

1. The state of being operable used in availability generally implies that an inoperable item can be restored to an operable condition by maintenance activities.
2. The need for an item to be committable when required implies satisfactory, reliable operation.

It can be shown that, in a steady state condition, availability can be expressed as

$$A = \frac{MTBF}{MTBF + MTTR}$$
(3.20)

where

$$MTBF = \text{mean time between failure}$$

$$MTTR = \text{mean time to repair}$$

MTBF may be calculated for any interval by summing the total functioning life of the population and dividing this value by the total number of failures that occur during this interval. This definition holds for any measure of component or system life, such as time, distance, cycles, or events. *MTBF* and *MTTR* are related by the fact that the *MTBF* is the expected value of the *MTTR*. This is written

$$MTTR = E[MTBF].$$
(3.21)

MTTR is defined as the total time required for corrective maintenance, divided by the total number of corrective maintenance actions required during a given period of time. The value of an element's *MTTR* can be quite critical in relation to the safety event. For example, given the occurrence of a hazardous event, such as inability to supply oxygen to a human, a submarine, or an aircraft, it is obviously desirable that the value of *MTTR* be less than the time required for the event to cause injury or harm to the system.

3.6 INTERFACE WITH HUMAN FACTORS

As indicated earlier, one of the fundamental qualities of system safety is that it relates to and interfaces with both equipment and personnel. The interfaces discussed thus far in this chapter relate mainly to equipment. Within the last few years, however, considerable attention has been focused on the interface between safety and personnel activities, particularly on the incidence of human error. A survey of nine Air Force missile systems conducted by the Stanford Research Institute, for example, determined that human error contributed from 20 to 53 percent of system unreliability. A study by L. W. Rook, Jr., of the fabrication process of an Atomic Energy Commission contractor determined that approximately 82 percent of electronic production defects could be directly attributed to human error. The nuclear accident at Three Mile Island that caused concern about a core meltdown (see Section 2.15.2) has been attributed largely to human error.

Just as Figure 3-3 illustrates the notion that an unreliable sample point is not necessarily equivalent to an unsafe sample point, it does not follow that a human error must give rise to a hazardous sample point. Rather, the data developed by studies indicate that there is a need to identify the causes and reduce the number of human errors, for it must be presumed that some of these errors do give rise to hazardous events. Studies examining how personnel function and malfunction are part of a discipline known as *human factors* in the United States and *ergonomics* in other countries. This discipline may be defined as

> A body of scientific facts about human characteristics. The term covers all biomedical and psychosocial considerations. It includes, but is not limited to, principles and applications in the areas of human engineering, personnel selection, training, life support, job performance aids, and human performance evaluation.

All areas mentioned in this definition are assumed to be generally understood, with the possible exceptions of human engineering and human performance, which are defined as follows:

> *Human engineering* is the area of human factors which applies scientific knowledge to the design of items in order to achieve effective human-machine integration and utilization. *Human performance* is a measure of human functions and actions in a specified environment.

In relation to safety, the role of human factors engineering is to eliminate or control hazards at the interfaces between humans and machinery in *human-machine systems*. This is defined as

> An arrangement of humans and equipment used to carry out a mission or to obtain a system output.

Control of hazards at this interface is aided by considering basic human characteristics in the design of the system—muscular strength and coordination, body dimensions, perception and judgment, native and acquired skills, optimum and maximum workload, and the need for comfort and freedom from environmental stress. The first groups organized in this field attempted to fit machines to people, mainly through the interdisciplinary combination of psychology and engineering. In more recent times, as machines have grown larger, more complex, and more powerful, the discipline has shifted its point of view so that both human and machine are treated as a single entity, a human-machine system. Two problems confront the human factors engineer because, although one of the primary objectives of human factors is to enhance safety by relating equipment to human capabilities and limitations, much of the biological and psychological information needed for this purpose is not yet available and the mathematical tools for quantifying and optimizing human factors in a formal fashion need further development.

It is not intended—in fact, it is not possible—for the practitioner of system safety who relates to human-machine systems to become independently knowledgeable of psychology, physiology, sociology, and those other life sciences needed by human factors engineers to fulfill their responsibilities. Rather, it is suggested that the safety analyst be aware of the character of this responsibility in order to be able to assess if the effort carried out to fulfill the responsibility is adequate in scope and execution. Perhaps the most fundamental interface of human factors and system safety is the one relating to atypical environments. These include

1. Those mechanisms by which the body regulates and maintains an optimal internal environment. This ranges from thermal regulation, used by the body to defend itself against external heat and cold, to emotional need fulfillment techniques used by the body to cope with the tasks presented at the human-machine interface without introducing errors of omission or commission caused by emotional transients. This latter consideration is relevant, for example, to automotive safety, for which it is interesting to speculate about how many of the 60,000 automobile accident deaths that occur annually could be prevented by greater knowledge of emotional factors in the work environment. Studies of personality characteristics of professional drivers, in which parental history, neurotic traits in childhood, school adjustment, employment and armed service records, and sexual and social adjustment were considered, indicate significantly higher accident rates for those who had emotional problems in personal or social adjustments throughout their lives. One Los Angeles firm with a company-paid auto insurance plan has found that drivers rated as low-risk by the insurer miss work an average of only seven days a year while the average high-risk driver misses thirteen days a year and has to be reprimanded three times as often on the job.

2. Ability to adapt to specific work-sleep schedules while maintaining a high degree of effectiveness. Examples include the work-sleep schedules

needed by the military during operations; those established by the FAA and the airlines to control the amount of rest required for airline flight crews between flights; and the circadian calendar that controls people and, consequently, can induce inefficiencies in an individual whisked by a fast-flying jet into a time zone that is different by several hours from the individual's customary one.

3. Human tolerance to atypical forces such as acceleration, shock, vibration, and noise. These are increasing in our environment as we increase the amount of time spent riding in various transportation devices, using mechanical, electrical, and pneumatic tools at home and at work, and being surrounded by the noise of others engaging in such activities. This category also includes hazards that result from too great a reduction in forces usually present, such as the effects of long-term zero gravity, for example. Finally, the desirability of introducing an occasional force into a human-machine system needs to be considered sometimes. For example, it has been shown that accidents are reduced by projections that extend above the roadbed, serving as markers and dividers between lanes on multilane, high-speed highways and introducing thumping and noise during intentional lane changing or unintentional meandering.

4. Human tolerance to long-term effects of irreversible, or slowly reversible, pollutants emitted into the environment, such as radioactivity, industrial pollutants, sewage, insecticides, herbicides, and other chemicals. Perhaps the most profound aspect of this problem is the inability to describe observed changes as a function of long-term, low-level inputs in a precise, quantitative fashion.

Another interface of safety and human factors is the space in which people work, live, and play, including the space itself, the furniture within the space, and any special equipment used to ease, amplify, and help implement individuals' movements, perception, and ability to respond to routine and emergency conditions. These conditions are equally applicable to aquanauts, cosmonauts, and homemakers, all of whom are surrounded by helpful but potentially hazardous implements. The following are of special interest in this area, perhaps because of their dramatic character:

1. Sensory supplements intended to enhance, amplify, or transmit data perceivable without aid or designed to transform data into the electromagnetic, auditory, or tactile range capable of being perceived by humans. These may include instruments intended for use by people with limited or curtailed senses as well as by normal individuals.

2. Remote manipulators intended to extend or amplify physical abilities. These include devices, such as microsurgical instruments, that add strength or refine or permit the performance of more delicate work.

3. Adaptive systems that are capable of recognizing patterns, learning, and making changes in their outputs in response to changing inputs.

Clothing may be considered a special case of the space about people that needs to be designed according to both human factors and safety considerations. The suits worn by the astronauts for walking on the moon illustrate an esoteric application; on a more mundane level, consider the hundreds of variations in gloves used in industry, around the home, and by hobbyists.

As in the case of system safety, human factors engineers are intensively involved in quantifying and optimizing their discipline and in relating it to the larger domain of system effectiveness. Interestingly, one of the major tools employed by both in this quantification is the fault tree, which is discussed in Chapters 6 to 8.

3.7 *INTERFACE WITH QUALITY ASSURANCE*

The interface between quality assurance and system safety occurs primarily in relation to the quality assurance efforts that are intended to prevent hazards caused by defects in material and workmanship. *Quality assurance* may be defined as

> The planned, systematic set of actions taken to provide adequate confidence that a system will perform satisfactorily when put to use.

The phrase "adequate confidence" is relative rather than absolute, varying from system to system. Ideally, adequate confidence is quantified by means of an optimization process. The role that quality assurance plays in relation to a system during this process is to ensure that all the characteristics inherent in the design of a system remain invariant throughout manufacturing, shipping, storage, and normal use. The interface between a quality function and system safety is the part of the effort that provides assurance that the safety level remains invariant and that new hazards are not introduced into the system, for example by maintenance activities. The primary mechanism by which the quality function provides this assurance at the interface is inspection of a set of quality characteristics jointly determined by system safety and quality assurance to be appropriate for this purpose. A *quality characteristic* may be defined as

> Any physical or chemical property, such as a dimension, a temperature, a pressure, or any other parameter used to specify the nature of the system or a function the system is intended to provide.

For example, some of the quality characteristics that could be used to ensure that the design of a metal sphere remains unchanged through the manufacturing process include the radius of the sphere, the type of metal used, the hardness and grain size of the material, and the nature and extent of chemical impurities allowed in the product. Figure 3-7 presents the sequence of activities that revolve about a quality characteristic. The starting point occurs at the point marked "design," when the design is first specified. Proceeding clockwise in the figure, the next steps express the characteristic in terms usually incorporated in a specification and indicate the processes by which the quality characteristic becomes transformed into hardware. Selection of an improper process can make it difficult to achieve the tolerances specified for the design characteristic.

The first three aspects of quality characteristics shown in Figure 3-7— design, specification, and process engineering—yield software results. That is, the output of the effort is documentation. This resultant software is trans-formed into hardware by the next steps shown in the cycle. The step marked "purchasing" relates to the effort carried out during the interval between preparation of a purchase order and receipt of the material. The first inspection process describes work done by the quality assurance function to

Figure 3-7
Life Cycle of a Quality Characteristic

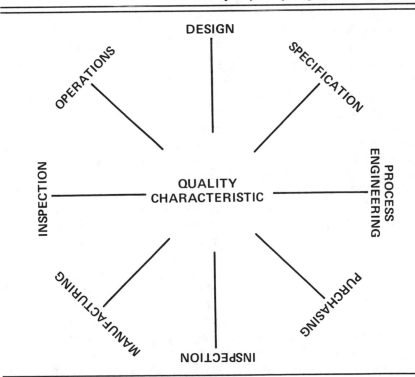

ensure that the characteristics included as a part of the purchase specification are incorporated in the material received. The inspection process conducted after manufacturing is intended to ensure that the desirable qualities existing at the time of receipt are not degraded by the manufacturing process.

The extent to which a quality assurance function is successful in preventing degradation of the safety level designed into a system can be estimated from data gathered during operations. If it is determined during operations that the inherent safety level has not been achieved in the system, further design effort is initiated, thereby starting the cycle again. The speed with which the design cycle is restarted varies in accordance with the differences between the inherent safety level and the safety level achieved, with rapid recycling occurring for large differences. There have been instances in the field of consumer goods (automobiles, for example) in which failure to respond rapidly has not only permitted hazards to remain in the product but has also been costly to the manufacturer in sales lost subsequent to the change because of a lingering bad reputation acquired in the interim.

In establishing the sample point and events—the set of quality characteristics which make up the interface between quality assurance and system safety—it is necessary to distinguish between "quality of design" and "quality of conformance." The relationships of value and cost for levels of quality of conformance and quality of design that fall below or above the optimum level are presented in Figure 3-8.

Quality of design (grade rating) is a function of the specifications that describe the system. For example, a Rolls Royce and a Chevrolet are both intended to provide the same function; they differ, however, in quality of design. Outside the quality assurance–system safety interface, quality of design is optimized by marketplace consideration. An improvement in quality of design that brings the quality level above the optimum costs more to achieve than can be returned in the marketplace. On the other hand, decrease in quality of design to a level below the optimum reduces costs by an amount less than the decrease that occurs in the value of the system.

Quality of conformance relates to the accuracy with which a system corresponds to its design criteria. A number of systems manufactured in accordance with a given set of specifications have the same quality of design but vary in quality of conformance, depending upon the accuracy and extent of the inspection process. Here too it is possible to derive an optimum based solely on monetary considerations outside the quality assurance–system safety interface. A level of quality of conformance that is above the optimum reduces the number of systems that require repair, maintenance, or replacement. However, the costs incurred in obtaining an above-optimum quality of conformance may be greater than the costs expended in changing the inspection process to bring about this improvement. The amount of money saved by decreasing the inspection process, resulting in a below-optimum quality

Figure 3-8
Costs for Other Than Optimum Quality

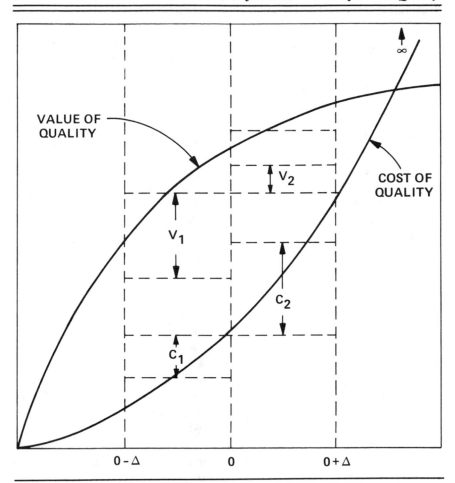

of conformance, may be less than the moneys required for carrying out the additional repairs, maintenance, and replacement that will then be required.

In Figure 3-8 the overall quality at level 0 is optimum, that is, it is most economical. A decrease in quality to the lower level, $0 - \Delta$, reduces the cost by C_1 but reduces the value by V_1, which is a greater reduction than C_1. An increase in the quality level to $0 + \Delta$ increases the value of the system by V_2, but requires a cost increase of C_2, which is a larger increase than V_2 to achieve this effect. In this instance the costs and values selected for analysis are oriented solely toward monetary considerations. Expansion of the analysis to include intangible values, the relative values of use and esteem noted in Section 2.10, is elaborated on in the next section.

3.8 INTERFACE WITH VALUE ENGINEERING

The notion that safety can be associated with a tangible value was introduced in Section 2.10. In particular, it was noted that a complete safety analysis often involves transforming relative safety values into absolute ones. An analogous problem exists in other fields. In economics, for example, considerable research has been carried out in an attempt to relate relative values to absolute ones by mathematical techniques based upon the laws of supply and demand and upon the principle of diminishing returns. The reader interested in the analogy between this aspect of economics and the interface between value engineering and system safety is referred to in Fisher's translation, *Researches into the Mathematical Principles of the Theory of Wealth* (see Additional Reading), a work published in French in 1839 by Augustin Cournot. This work is considered by many to be the first systematic exploration of the relation between relative and absolute values.

3.8.1 The Need for Absolute Value in Safety Analysis

Including relative value as a factor to be quantified can present some problems. Consider the complexity that would be added to Figure 3-8 if, instead of restricting value considerations solely to the costs required for repairing, maintaining, or replacing hardware, it were permitted to include injury or death. Needless to say, there is no convenient way to assess the value of a serious injury, let alone the value of a lost life, in monetary terms.

Nevertheless, in manufacturing or operating a system that is characterized by inherent hazards that can give rise to irreversible injuries or death, there is no recourse but to spend some amount of money to determine what hazards exist and to make arrangements to defray expenses incurred as a consequence of their occurrence. Experience has shown this to be necessary because injury or loss of life as a result of hazardous occurrences is usually transformed into a dollar value by court proceedings. Passing these costs on to the customer as part of the manufacturing and operating costs of the system can often hinder the competitive stature of a company.

This suggests that there is some optimum amount of money that should be spent for hazard identification and elimination or control in advance of system release for public use. The alternative of spending no money for this purpose and just buying liability insurance can be considerably more expensive. A well-known example of this problem is the tangle of litigation concerning the Ford Pinto. Not counting the costs associated with recalls, the costs awarded to customers by the courts were considerable. The largest single award (subsequently reduced to $6 million) was $128 million, and other large awards were $11.5 million and $3.1 million. To compound the incident, the company was indicted on criminal charges. Another example concerns Firestone, which estimated the cost of recalling controversial steel-belted radial tires at more than $200 million. These amounts appear minor compared to a $1 billion class-action suit filed in October 1978 on behalf of 5000 shipyard workers, charging 15 major asbestos manufacturers with conspiracy to conceal and to distort reports on the health damages of asbestos.

The number of lawsuits brought to court, many of them successfully, against the medical and drug professions for malpractice or damages illustrates how far-reaching the consequences of product hazards can be. The average amount of money awarded per case as a result of such lawsuits has risen considerably in recent years. Damages of $2.2 million were awarded in June 1971 to a California mother and her daughter as a settlement for injuries to the child resulting from the effects of the drug thalidomide taken by the mother during the first trimester of pregnancy.

This threat of lawsuits has resulted in an increase in product costs for additional testing and complex record keeping for potential recalls, for example, that is passed on to customers. An analogy exists in services. Some people claim that fear of malpractice suits has caused some members of the medical profession to treat patients in a more defensive fashion. This not only results in increased costs but can also add future health risks to patients if excessive precautionary procedures such as X rays are used for the doctors' protection.

Suppose that the entire medical profession, a hospital, or a single physician is thought of as a system in the sense defined in Section 2.6. Suppose further that this system is treated in accordance with the value and risk criteria suggested in Sections 2.10 to 2.13. It becomes apparent then that there is some hazard level greater than zero that must be incurred by both users and operators of the system. The claim that a posture of increased "defensiveness" may lower the number of malpractice suits and, consequently, the cost to the patient needs to consider the distributed costs, monetary and otherwise, borne by all patients treated "defensively." One difficulty that arises in optimizing this system is the unpalatability of accepting risks when seeking medical treatment.

3.8.2 Analysis for Absolute Values

A value analysis oriented toward the notion of elimination or control of hazards is presented in Figure 3-9. The significant characteristic of this figure is that some explicit dollar value is assigned to the system for each significant inherent hazard known to exist. This implies that all relative values that affect

Figure 3-9
Value Analysis of Hazard Elimination

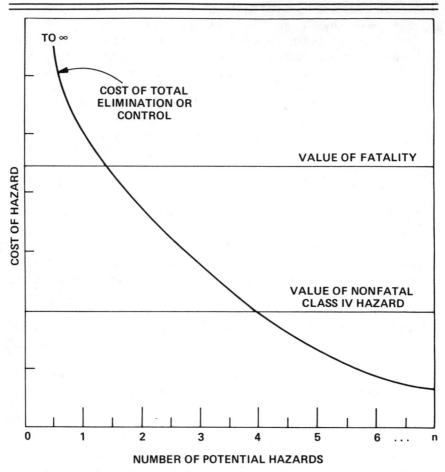

the inputs or outputs of a system have been transformed into absolute values. A mechanism for achieving this is introduced in the next section.

Two composite values, one for fatal occurrence and one for nonfatal hazards, are shown in Figure 3-9. It can be seen from the figure that the cost of eliminating a single hazard is relatively small when there are a large number of hazards inherent in the system but that the cost becomes relatively large as the number of inherent hazards remaining in the system approaches zero. The figure also shows that it would require an infinite amount of money to eliminate all hazards. The actual course to be followed with a given system in deciding what balance to attempt in a specific circumstance is a management decision. Management is aided in making this decision by having an assessment of the value of safety in absolute terms, that is, its cost.

3.8.3 Transformation from Relative to Absolute Value

One method for transforming relative values into absolute terms makes use of game theory. The relationship between game theory and the value of safety uses the notion of "utility" because of its widespread use in the literature in association with value. In general, particularly in the area of economics, utility is synonymous with value. In game theory, utility is used as a variant of the term *payoff*, and both utility and payoff are synonymous with E_o in this case. However, in game theory consideration is given to the value of the payoff in a manner that tends to maintain the equivalence of the terms *value* and *utility* as they are used in this text. A game consists of:

1. Players who have full knowledge of the rules of the game and the payoff functions for all players. Each player is assumed to be rational in the sense that, given two alternatives, the player will choose the one whose payoff has the greater value.
2. Strategies, one for each player.
3. Payoff functions, one for each player, the values of which depend upon the choices of strategy of all players.

In "playing" the game, each player attempts to maximize payoff where the outcome depends on the choices of the other players as well as on the individual's own choice. The selection of a strategy by each player depends upon an estimate of the value of the payoff, and in general the value of the payoff to each player depends upon both absolute and subjective considerations.

To relate payoffs to utility, consider a game with payoffs a, b, and c equal to 0.9, 0.7, and 0.2 units, respectively. Let the numbers represent nondimensional probabilities relating to the likelihoods that a given hazardous event will occur in accordance with the strategy selected. The intervals between these payoffs are 0.2 and 0.5, that is, $(0.9 - 0.7) = 0.2$ and $(0.7 - 0.2) = 0.5$.

A payoff is called a "utility of the interval scale," and it is required that the payoff be invariant under transformation. As an illustration of one such transformation, notice that adding 0.05 to each of the payoffs a, b, and c does not change the intervals defined by $(a-b)$ and $(b-c)$. Although this illustrative transformation does not alter the utility of the payoff there is a decrease in value in the sense that the likelihood of the hazardous occurrence has been increased regardless of the strategy selected. Game theory, therefore, provides a procedure with standard rules for transforming relative values, values that are differently interpretable by different people, into absolute ones.

Additional Reading

Bates, J. A. Some characteristics of the human operator. *I.E.E. Journal*, 94, no. 2, Part IIA, (1947), pp. 298–304.

Bazovsky, I. *Reliability Theory and Practice*. Englewood Cliffs, N.J.: Prentice-Hall, 1961.

Charnes, A., W. W. Cooper, and B. Mellon. A model for optimizing production by reference to cost surrogates. *Econometrica* 23, no. 3 (July 1955), pp. 307–323.

Fisher, I., *Researches into the Mathematical Principles of the Theory of Wealth*. New York: Augustus M. Kelley, 1927. ISBN 0-678-00066-2.

Juran, J. M. *Quality Control Handbook*. 3rd ed. New York: McGraw-Hill, 1974.

Lingren, N. Human factors in engineering. *IEEE Spectrum*, March 1966, pp. 132–139.

Lloyd, D. K. and M. Lipow. *Reliability: Management, Methods and Mathematics*. 2d ed. Published by the authors, 1976.

Malasky, S. W. Value engineering aspects of system safety in manned space programs. *Journal of Value Engineering*, April 1967, pp. 169–175.

Morrison, T. A Measure of Effectiveness for Weapon System Evaluation. Report, the Operations Research Society of America Nineteenth National Meeting, May 25–26, 1961.

Ramo, S. *Cure for Chaos; Fresh Solutions to Social Problems Through the Systems Approach*. New York: McKay, 1969.

Sell, R. G. Ergonomics and safety. *Hazard Prevention*, November–December 1979, pp. 10–14.

University of New Mexico and Subcommittee of Human Factors of the Electronic Industries Association. *Proceedings of the Symposium on the Quantification of Human Performance*. August 16–19, 1964. E.I.A., Washington, D.C.

U.S., Air Force Systems Command. *Weapon System Effectiveness Industry Advisory Committee, Chairman's Final Report*. AFSC-TR-65-6, January 1965.

Chapter 4

PROGRAM FOR ACHIEVING SAFETY

4.1 THE PROBLEM-SOLVING PROCESS

Ideally, the process of establishing a program and carrying out the tasks necessary for achieving a given level of safety are conducted in accordance with the rules of scientific methodology introduced in 1620 by Sir Francis Bacon in his book *Novem Organum.* One of the tenets of this method establishes the need to use both deductive and inductive logic to derive results that are to be used for prediction. The essence of system safety is its predictive character.

In modern format, the application of scientific method for establishing and carrying out a program to achieve safety requires that five tasks be accomplished:

1. Formulate the problem, including the establishment of any limitations that may be necessary to make the problem manageable. For example, further refinements in research into the relationship between smoking and lung cancer might utilize a control group of individuals of a discrete age group and exclude pipe and cigar smokers.
2. Design an experiment or establish a data collection activity for proving or disproving hypotheses. This effort includes the task of determining sample sizes and the nature of data needed.
3. Carry out the planned experiments.
4. Tabulate and describe the results of the experiments. This effort includes the preparation of charts, graphs, and probability distributions, and performing calculations based on the data.
5. Formulate an answer or generate an inference. (The latter is often the only option available.) For example, attempts to quantify the relationship between smoking and lung cancer may require conclusions about all smokers based on data derived from segments of the smoking and cancer universes.

A method of presenting the above steps in graphic form and showing some of the interrelationships between them is introduced in Section 6.4.2 and illustrated in Figure 6-2.

4.1.1 Implementation Difficulties

Difficulties encountered in attempting to follow this procedure are brought about mainly by an inability to formulate the problem with sufficient explicitness or to implement the resultant solution obtained by the methodology. Difficulty in problem formulation is often caused by conflicting demands among the various parameters considered in the optimizing process. In some cases formulation is hampered because of the existence of unknown inherent hazards or because known hazards are incorrectly presumed to be of little consequence.

Other problems in implementing this procedure can occur at interfaces where various organizations have joint responsibility for one or more aspects of safety, the usual case. Assistance, consultation, guidance, and even checks and balances may be provided by a formal system safety organization, but true responsibility for ensuring that a satisfactory safety level will be achieved for the system belongs to those at the highest management level.

Since system safety is relatively new in its deliberate application and is to some extent a software activity, its function and method of operations are not always fully understood by everyone in top management. It may be hard for some management personnel to perceive the potential impact of system safety or to see its effect on a program. Consequently, in order to obtain the requisite management support for maximum effectiveness, the system safety program should be structured so that its impact on the system is clear. This implies the need for thoughtful and careful determination of the interface relationships between system safety and other organizations and tangible evidence of impact resulting from these relationships. Useful evidence includes documented changes to the design resulting from safety analyses and subsequent recommendations.

4.1.2 Organizations Collaborating with Safety

Figure 4-1 presents some of the various organizations that may interface with the system safety function and indicates the nature of some of the interfaces. Such interfaces are not necessarily limited to one political agency or one parent company. Rather, the safety organization that has primary responsibility for achieving the safety objectives established for the system must relate to the customer, associate business organizations, subcontractors, government agencies, and the ultimate users of the system as well as to organizations in the same agency or company.

Figure 4-1
System Safety Organizational Interfaces

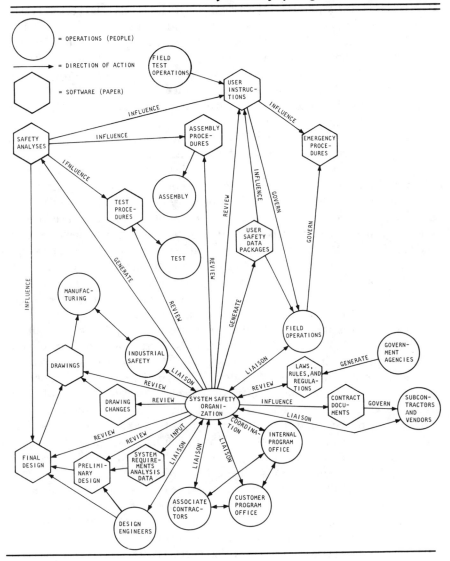

The safety-customer interface that occurs when the customer is also the user (rather than a case where the customer and the user are different) has interesting implications. In the former context, the term "customer" refers to the entity that contracts with the organization responsible for achieving the safety level of the system, as in the purchase of a turbine or nuclear generator by a utility company or the purchase of a passenger aircraft by a commercial airline. In these cases, the customer—the utility or the airline—can engage in dialogue directly with the system safety manager responsible for ensuring that the specified level of safety is achieved. If necessary, the customer can make modifications to the contract to adjust the safety level.

On the other hand, the ultimate consumer of equipments, such as home appliances, or services, such as medical tests performed by laboratories, is often not in a position to relate directly to the organization responsible for the system's safety. In these cases, government often interfaces on behalf of the user with the organization primarily responsible for assuring the safety of the system.

As noted earlier, there has been a recent increase in the amount of dialogue undertaken between the ultimate user and the developers of consumer systems (products), primarily through court action. Further, recent court decisions in this area suggest that the courts are placing requirements upon the system safety function to relate more with the safety needs of the ultimate user. This has resulted in concern about whether the allegiance of the safety analyst lies with the company for which he or she works or with the ultimate users. The requirements established by some companies that only professionally registered safety engineers be hired has furthered this polarization, because such registration includes a set of ethical guidelines sometimes felt to be in conflict with company requirements. On occasion such perceived differences have resulted in dramatic confrontations. Two notable examples are the safety debates that arose during development of the San Francisco Bay Area Rapid Transit (BART) system and the controls and personnel training in nuclear electricity plants. A corollary consequence has centered about the liability of the safety analyst in the event a hazard occurs, which affects the amount of insurance needed to settle potential court actions and raises further questions about whether this insurance should be provided by the company, the safety analyst, or be shared. This problem is of concern because of the possibility of long-term liability. In some instances the courts have awarded damages to individuals injured 25 years after product delivery. Suggestions are being offered at federal and state levels to limit the interval of time a manufacturer can be held responsible for hazards. It is estimated that more than 95 percent of product-related injuries occur within six years for consumer products and that more than 80 percent occur within ten years for heavy industrial products. However, consumer organizations are protesting arbitrary cutoff points for lawsuits.

4.2 SYSTEM SAFETY PROGRAM PLAN

One of the primary aids for carrying out the problem-solving procedure described in Section 4.1 is the System Safety Program Plan (SSPP). Such a document can govern all the safety effort carried out during the life of a program. A complete SSPP contains the following elements:

Figure 4-2
Life Cycle of a System

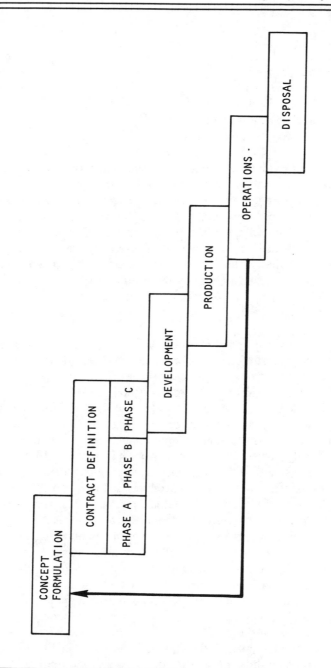

1. Planning
2. Organizing
3. Contracting
4. Interface coordination
5. Establishing criteria
6. Performing analysis
7. Evaluating
8. Data retention and updating
9. Reporting progress and results

An outline for a generalized SSPP which shows how these nine elements are integrated over the system life cycle is presented in Appendix C. In preparing an SSPP for a given system, it is necessary to determine which portions of this outline may be deleted and which require further amplification. This determination is based upon the nature of the system, the requirements established for its use, and the role that the responsible safety organization has in relation to the system. The next two sections elaborate on two aspects of a generalized program plan that are almost always applicable.

4.2.1 Time Span

The time span covered by the program plan is controlled by the life cycle of the system. A general life cycle is presented in schematic fashion in Figure 4-2. The overlap shown in Figure 4-2 between the various phases of the life cycle will, of course, vary from system to system and, for any given system, may be a function of the number of systems produced. The disposal phase, although relatively new as a part of the system life cycle, is now widely incorporated in system planning. Some applications are exotic, concerned for example with nuclear waste or with orbiting spacecraft that may reenter the earth's atmosphere. Less exotic perhaps but of considerable importance are concerns for after-use disposal of products such as insecticides, herbicides, industrial wastes, and nonbiodegradable containers. The several subphases shown in the contract definition phase represent the general case. There may be only one phase. When there is more than one subphase, there may be some overlap between them. The feedback of information from operations to concept formulation can occur over one of several possible feedback loops. There may be other inner loops in the system life cycle, such as between the production and the research and development phases or between the research and development and contract definition phases.

An overview of the relationship between the SSPP and its various elements and the life cycle of the system is presented in Figure 4-3. In this figure, the area contained within the outside rectangle describes the program established for the system, with time increasing clockwise. The four major

Figure 4-3
System Safety Program Plan Interfaces

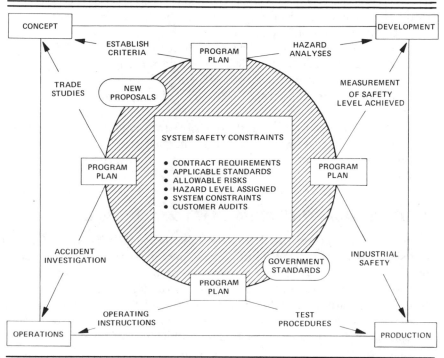

phases of a system life cycle are denoted by the rectangles at the corners of the figure, beginning at the upper left with "concept." The circle in the center of the figure represents the portions of the system life cycle activities that are governed or strongly influenced by the SSPP. Some major constraints on the program plan which establish the scope of the system safety activities are noted in the center square. The two semielliptical figures represent external sources of influence on safety and the program plan.

4.2.2 Legal Character of the Plan

The control and influence that an SSPP has upon program activities are usually grounded legally. Legality is established by making the plan part of a procurement contract which provides a general description of the system. Consequently, the set of tasks established by an SSPP are directly traceable to requirements in the contract. In some instances tasks established by a plan are intended to achieve objectives that are more stringent than requirements established by the contract. Circumstances in which this may be desirable are discussed in the next section.

In order for a program plan to be a working engineering document rather than a philosophical treatise, it must be specific in content. It must describe "how," "when," and "what" for the tasks described in the plan. Procedures describing "how," "when," and "what" that are too detailed to include in the plan can be provided in the separate, supporting documents, provided these documents are identified in the plan. Some general comments about the various portions of the SSPP outlined in Appendix C are presented in the next few sections. The material deals with those aspects of the program plan that may be directly traceable to a contract and with those portions of the program plan that are intended to provide guidance by establishing goals and objectives which are more stringent than those established by contract.

4.3 INTRODUCTORY PROGRAM PLAN MATERIAL

This section is concerned with the four units that make up item 1, "General Considerations," of the SSPP outline presented in Appendix C. This section of the plan is the most general portion and permits some exposition about system safety and its relation to the system under consideration. In this section the relationship between the safety level to be achieved and the amount of effort to be expended is considered, the philosophy of the risk acceptance is discussed, and the goals and objectives of the safety program are outlined.

The general section may also contain, usually under the heading of "scope," a description of the requirements, qualitative and quantitative, that are to be achieved as a result of implementing the program plan. In addition, this section may contain any noncontractual goals and objectives that are established for the program. If, for example, there are severe penalties, either monetary or in potential loss of business in the event of a failure to achieve a required safety level or accident rate, it may be desirable to establish a higher safety level as an objective in order to provide a "safety" margin for the effort to be conducted. In other instances the risk of potential lawsuits brought by the ultimate consumer may make it desirable for a manufacturer to establish a safety goal that is higher than that established by formal requirements. Formal requirements for consumer products are usually governed by industry standards or government regulations. Recent Supreme Court decisions have tended to increase manufacturers' responsibilities for accidents occurring to the ultimate user that can be judged to be a result of design deficiencies. (The applicable time limit for such deficiencies is discussed in Section 4.1.2.) In addition, the definition of design qualities that may be considered deficiencies has been broadened by court decisions.

The section on applicable documents represents one of the more formal sections of the System Safety Program Plan. Documents incorporated in this portion of the plan may include

1. Company standards, policies, and procedures, including those that may be required by higher organizational authorities, such as those promulgated at a corporate level.
2. Specifications and standards established by government agencies that have authority over the procurement process or that control the operation of the system. Examples of such agencies include the Department of Defense, the Federal Aviation Authority, and the departments concerned with health, education, and welfare. Specifications and standards established by such agencies may exercise controls at levels ranging from the entire system to its smallest component. In addition, control may be exercised over the qualifications of the people who construct and operate the system, the procedures governing the manner in which the system is operated, and the conditions in which the system may be used.
3. Safety regulations established by local, state, and federal governments. Sometimes different political entities vary in their treatment of identical hazards. Consequently, a system which crosses political boundaries during its lifetime may find itself relating to inconsistent regulations.

In specifying a set of documents for inclusion in a program plan that is applicable to a given safety program, it is necessary to distinguish between those documents that require adherence in a formal, legal fashion and those that are included for guidance. Since the term "guidance" is open to interpretation, it is preferable that a program plan establish without equivocation the extent to which each document included in the plan is applicable to the system.

4.4 ORGANIZING FOR SYSTEM SAFETY

This section discusses item 2, "Safety Organization, Responsibilities, and Authority," of Appendix C. Theoretically, it is possible to describe a unique system safety organization that is optimal for carrying out a safety program suitable for any given project. From a practical point of view, such an organizational entity can rarely be described. However, all of the tasks, including those in the domain of safety, that need to be conducted so that a given system may be conceived, designed, manufactured, and used can usually be specified. There is usually a variety of ways to organize so that these tasks can be accomplished, although some organizational entities intended to accomplish a specific task are inherently more effective than others. In some cases, the effectiveness of an organization depends upon some key people, a difficult factor to include in an organization chart.

Nevertheless, some organizational guidelines may be offered which are relatively invariant in relation to the type of system under consideration and the tasks that need to be carried out. To begin, a *system safety organization* may be defined as

A group of people responsible for establishing, managing, and performing an overall system safety program.

This definition places no constraints on the size or character of the safety organization. Furthermore, the definition is equally applicable to system safety organizations that are readily identifiable as a single entity and to those that are diffused over an entire project or company and cannot be represented on a conventional organization chart as a single entity. If the system safety organization is diffused, it may be convenient to define the *managing activity*. This is

The organizational element that plans, organizes, directs, and controls tasks and associated functions throughout one or more of the life-cycle phases of the system.

The managing organization can be wholly contained within the system safety block on an organization chart, or it may be distributed between the system safety block and program management block.

Although one cannot always prescribe a unique system safety organization, it is possible to infer the importance of a safety function to top management by examining two primary factors, budget size and reporting level. These two are rarely independent, for a high level of expenditure for safety invariably implies management attention. Conversely, the inescapable conclusion to be drawn about a program which has a proportionately small budget for safety and either no individual explicitly responsible for this effort or responsible personnel too low in the management hierarchy to be effective is that top management considers safety to be relatively unimportant. The correctness of management's choice of the relative importance of system safety to other organizational elements can only be assessed if all safety implications of the system are known.

In any event, the System Safety Program Plan needs to contain a precise description of the organization of people responsible for establishing, managing, and performing the overall system safety program. In some instances, particularly if the plan is part of a solicitation for new business, it may also be necessary to provide the names of the individuals in the safety organization along with a description of their qualifications. In general, the plan needs to include

1. The functions and responsibilities of all persons associated with establishing policy or implementing any aspects of the safety program.
2. Details about the authority delegated to each individual in the safety organization and the relationships among line, staff, interdepartmental, project, functional, and general management organizations.

A similar organizational description, along with a set of functions and responsibilities, is needed for any committees or boards that are discussed in the program plan where their functions are beyond the scope of those established for the system safety organization. For example, the SSPP developed by a company for controlling its safety efforts may include a requirement that the company provide representation to a corporate level safety staff. This presumes, as is often the case, that the corporation establishes safety policies and provides guidance to the various companies and divisions through the use of a safety board whose composition includes representation from each company. In some instances, corporate policy prevents a division from engaging in activities considered to be unusually hazardous without explicit approval from the corporate office. The safety board, acting as advisor to the corporate office, may be responsible for conducting the analysis upon which the decision is based about whether to allow a member company to proceed with a hazardous business activity.

Another illustration is based upon an example of a consortium of associate contractors that is vested with responsibility for developing or operating a system. In this case, it is reasonable to presume that the set of system safety program plans intended to control all associate contractors will provide for a board to control or guide the overall safety program and require representation by each associate. The board can be expected to provide policy and guidance for the overall program and to exercise control of those hazards that may cross the interface between equipments assigned to individual members of the consortium.

4.5 SYSTEM SAFETY CRITERIA

The outline of item 3 of Appendix C, "System Safety Criteria," is essentially self-explanatory, and, to a certain extent, its contents are optional. For example, definitions can be included for the basic terms that are used in the SSPP or a glossary can be provided that is wider in scope and defines all of the terms in the safety domain relating to the system. Similarly, the section that identifies those analytical techniques to be used can range from a list to a complete description of the methodologies and the mathematical basis upon which each methodology is founded.

The subsection concerned with special contractual requirements is of particular significance because these terms may exercise a considerable influence on the scope and magnitude of the effort to be conducted. Special contractual requirements can range from the presentation of trading stamps to employees for exceeding a given number of accident-free days to contract cancellation as a result of lost-time accidents caused by use of the system. There may be some overlap in content between the subjects listed in item 3 and those listed in item 1 of Appendix C. For example, "Special Contractual Requirements," item 3.4, might be introduced as a part of item 1.2, "Scope and Purpose."

A distinction should be made as shown in the appendix between "routine" and "special" considerations, if any exist, for use in two types of situations because each is handled more efficiently by program plans of different formats. When the program plan is intended to relate to one given system or several identical systems, there is no advantage in attempting to distinguish between routine and special contractual considerations. However, for the case in which a program plan must relate to a variety of systems, it may be convenient to use separate plans. These could consist of one general purpose program plan plus addenda oriented toward each individual system. In this configuration, the general purpose plan is used to describe the requirements implied by item 1.2 of Appendix C, "Scope and Purpose," that are common to all systems. Each member of the set of specific system plans is then used to relate uniquely to one specific system. These requirements are implied by item 3.3 of Appendix C, "System Safety Precedents."

Another instance in which it may be more efficient to distinguish between routine and special contractual requirements occurs when one complex system is used to carry out many functions and consequently interfaces with other systems. Examples of this category include factories with multi-product lines, hospitals with several specialized departments, and transportation systems utilizing several types of vehicles for transporting different mixes of products and people. Here, too, it may be effective to establish one general purpose program plan which, with appropriate addenda, controls the safety of the several functions the system is called upon to perform.

4.6 SYSTEM SAFETY TASKS

The material in this section discusses item 4 of Appendix C, "System Safety Tasks to Be Completed." First, it should be noted that there can be no rigid assignment of safety tasks for the various phases of the system life cycle that are presented in Figure 4-2. However, the tasks required of system safety that need to be carried out to fulfill the safety objectives of the system can be divided into three general groups and can be roughly assigned to specific phases of the system life cycle. Figure 4-4 presents these three groups—safety analysis, technical data evaluation, and support—and indicates the fundamental efforts that fall within each group. It is to be expected that some of the tasks performed will not be completed within the life cycle phase in which they are started. The time when each task should be initiated is, however, generally assignable to a specific phase of the system life cycle.

The tasks considered to be fundamental for conducting a system safety program for a complex system are discussed in the next few sections in an order corresponding to the system life-cycle phase in which each task is normally initiated. In establishing a set of basic tasks to be carried out for a system of small or moderate complexity or for a system safety effort which is concerned with only a portion of a complex system it may be appropriate to delete some of these tasks or to limit their scope. It should be emphasized that the sets of tasks presented in the next few sections do not define in detail the entire scope of a system safety program for every system. Some comments on the documentation needed for a complete system safety program plan are presented in Section 4.8.

4.6.1 Concept Formulation Phase

1. The technical approaches that may be applied as system safety considerations for the various design features are evaluated at this time. In addition, trade studies necessary to select the optimum approach from among the various possible considerations are initiated in this phase.
2. The task of identifying potential safety interface problems is initiated. This is particularly applicable to systems that are composites of equipments designed and manufactured by more than one contractor. Suppose, for example, that a subsystem being developed by an associate contractor for integration with other subsystems developed by other associates contains ordnance or flammable materials. In that case it would be necessary to make any interface problems associated with such materials known to the other associates whose equipment or procedures have the potential for providing sources of ignition. A "central" system safety organization might be called upon to perform this function.
3. Those areas that require safety investigations because they are novel or are extensions of the state of the art are identified in this phase.
4. The system safety performance envelope parameters, such as maximum permissible accident rate or number of allowable fatalities, are now defined. This task can have a profound effect upon the design and testing of the system.

Figure 4-4
System Safety Functions

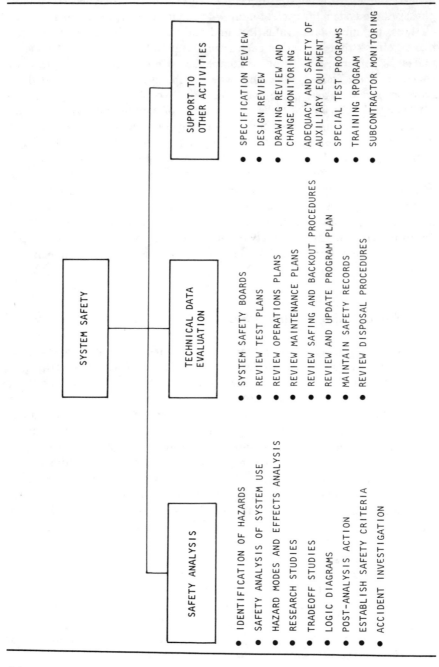

5. Any areas that require special consideration by safety, such as the need to establish new safety rating requirements for the system, limitations upon the use of the system, or the number and types of allowable risks to be taken during testing of the system, should be identified at this time.

4.6.2 Contract Definition Phase

The contract definition portion of the life cycle presented in Figure 4-2 is shown as having three phases. The distinction among these phases, assuming that there are exactly three, is presumed to be as follows:

1. The procuring agency, government or private, issues a request for proposal to a group of contractors selected by some scheme during phase A. The contractors determine during this phase whether to respond to the request and, if they choose to respond, establish an organization responsible for developing the response.
2. A response to the request for proposal is prepared by each contractor during phase B and forwarded to the procuring agency.
3. During phase C the procuring agency evaluates the various proposals submitted and determines which of the contractors are to receive awards. Activities conducted by a contractor during this phase of the life cycle are carried out on a risk basis, since there is no certainty of receiving an award. Formal initiation of work is considered to occur at $t = 0$ in the development phase of the system life cycle.

The bulk of system safety tasks conducted during phases A and C are oriented toward procurement, while the main effort carried out during phase B is oriented toward defining and initiating those tasks needed for the development, production, and operations phases of the system life cycle.

The system safety tasks discussed next are those that are usually initiated during the contract definition phase of a system life cycle. No attempt is made in the discussion to relate these tasks explicitly to any of the three segments of the contract definition phase, and no distinction is made for these tasks between procurement and implementation orientations. In general, these tasks are as follows:

1. A preliminary draft of the system safety program plan is prepared during the early part of this portion of the system life cycle. The nature and purpose of the SSPP is discussed in Section 4.2. Although it is a task to be conducted by the system safety organization that is begun early in the life of the system, the plan is updated on a regular basis throughout the entire life of the program.

2. A hazard analysis, the kernel of system safety activities, is initiated during the concept formulation phase. If not, the hazard analysis is initiated during this phase is developed further in the contract definition phase. The purpose of the hazard analysis is to determine what hazards are present in the system, the effects of their occurrence, how they may be prevented or controlled, and to assess the risk associated in using or operating the system. Procedures for carrying out a hazard analysis are dicussed at some length in succeeding chapters.
3. The system safety requirements and constraints that need to be included in higher-level system specifications are now prepared.
4. The data required from all sources—internal and external—with which the system safety organization will interface upon award of the contract are identified in this part of the system life cycle.
5. The safety content of the work statements to be included in subcontracts that need to be let shortly after award of the prime contract is prepared at this time. In particular, this task is conducted during the contract definition phase of the life cycle for those cases where delay in issuing statements of work after the contract award can prevent schedules from being met. This can happen, for example, when the system includes items that require a long lead time for delivery.

4.6.3 Development Phase

Initiation of the development phase of a system life cycle was defined in the previous section as the formal starting point of a program. Some of the tasks described as suitable for initiation in the contract definition phase might be better suited for initiation after award of the contract. If it is presumed, however, that the safety effort has proceeded as outlined thus far, then only those tasks identified in the program plan as having starting dates in the development phase need now be initiated:

1. Furnishing design criteria that are needed to establish the safety objectives of the various elements of the system and that, if achieved, permit the system objectives to be attained.
2. Evaluating the safety aspects of the system and its elements as it progresses through the various phases of the design. This is conventionally carried out as part of the review already programmed at specific milestones. (See Section 4.8.)
3. Preparing inputs to the various training courses intended to instruct operators and users of the system in the emergency safety provisions that are provided by the system.
4. Reviewing test plans to ensure that the safety features of the design are adequately tested.

5. If applicable, a trouble and failure reporting system should be developed at this time. This establishes a procedure for describing symptoms of problems as well as hazardous occurrences. Analysis of the former can be used to initiate preventive action to preclude actual occurrences.

In addition, most of the tasks initiated during earlier phases are updated and amplified during the development phase of the life cycle.

4.6.4 Production Phase

Comparatively speaking, few tasks are initiated by system safety during the production phase of the system life cycle. Safety tasks conducted during this phase are oriented mainly toward ensuring that the level of safety achieved during earlier phases is maintained during this phase. This is accomplished primarily through

1. Identifying critical manufacturing processes, inspections, and tests that could be degrading to the safety level of the system.
2. Establishing characteristics to be monitored by quality assurance to determine whether the safety level achieved earlier has been degraded.
3. Auditing engineering changes that are implemented during this phase of the life cycle to ensure that the safety level of the resulting design will equal or exceed the safety level previously achieved.

4.6.5 Operational Phase

Initiation of the operational phase of the system life cycle may be considered to occur when the customer accepts the first system for use. System safety tasks conducted during this phase depend to a large measure on the safety level achieved to this point and the nature of the contract. A low safety level in combination with a strong customer-user warranty clause will tend to ensure a high level of safety effort during this phase. In some instances the manufacturer-customer interface is extended through the operations phase by a trouble-reporting procedure which makes the manufacturer responsible for analyzing problems that arise after the system is delivered. Tasks generally conducted during this phase include:

1. Evaluations of any hardware and procedural changes implemented during this phase of the life cycle to ensure that the safety level achieved earlier is not degraded.
2. Review of operational activities to ensure that maintenance procedures carried out during this phase of the life cycle are not hazardous in themselves, do not degrade the inherent safety level, and do not cause

hazards that are external to the system. To ensure the latter apsect, it may be necessary to augment the hazard analysis to identify those problems that may be caused by maintenance activities. Further, it may be necessary to establish appropriate quality characteristics for use in the inspection process which is performed subsequent to maintenance activity to ensure that no hazard has been incorporated into the system as a consequence of the maintenance.

3. Evaluation of emergency procedures and any training program used to ensure that these procedures are carried out properly to establish the necessary confidence in their effectiveness.

4. Investigation of any accidents and incidents that may occur during the operational phase of the system life cycle. This task may include review of malfunctions and symptoms of potential problems, such as might be reported in a trouble and failure report. In some circumstances, boards of inquiry are established to identify the cause of accidents, develop procedures to be taken on an interim basis to prevent similar accidents from occurring, and determine design modifications appropriate for permanent elimination or control of the hazard. The system safety organization often acts as secretariat or chairs such boards of inquiry.

4.6.6 Disposal Phase

The procedures to be followed in disposing hazardous portions of a system— corrosive or toxic materials, nuclear wastes, easily ignited flammable matter, or ordnance items, for example—are best developed during early phases, primarily during the development phase. Two considerations are germane: (1) routine disposal at the normal end of system life, and (2) emergency disposal, a need that may occur at any time during the system life cycle.

The primary task required during this phase involves auditing the procedures used for disposal of the system's hazardous materials to ensure that they are being conducted properly.

4.7 GENERAL PROCEDURES FOR DEVELOPING SAFETY TASKS

Figure 4-5 represents a model for use in developing a set of tasks required by a system safety program plan for any phase of the system life cycle for any system. The starting point, a specified system objective shown at the top of the figure, describes the initial conditions and constraints upon which the safety tasks are to be based. The fundamental constraints upon safety is implied by the system objective, whose achievement requires completion of a specific set of tasks, including some for safety. There is also a safety objective to support the mission objective for which a group of safety tasks needs to be

Figure 4-5
Model for Developing Safety Tasks and Procedures

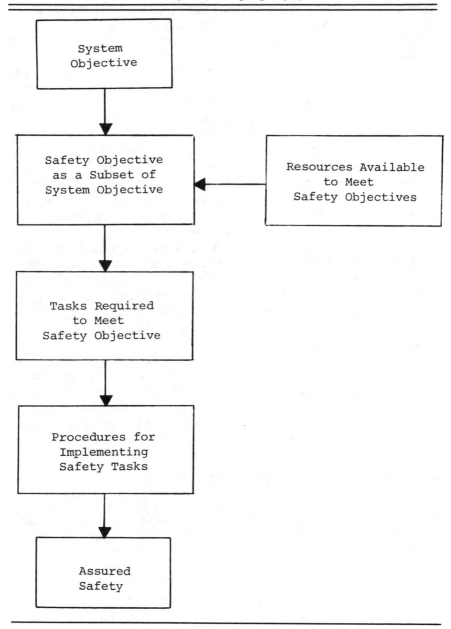

generated. The result of carrying out the safety tasks established to support the system and mission objectives is an achieved level of assured safety, obtained through the elimination or control of hazards shown at the bottom rectangle in the figure.

The first step in establishing a set of tasks that transforms a safety objective into an assured safety level is to describe the available resources, shown in Figure 4-5 by the right-hand rectangle. As an example, let the mission objective be to construct models of a new system in a factory currently producing other systems. Suppose one safety objective of the system is to minimize the number of industrial accidents that occur during the production of the new system. It is reasonable to presume that the system safety program plans prepared for the older systems describe a number of tasks applicable for achieving the safety objective for the new system. Therefore, it may only be necessary to prepare addenda to the program plans prepared for the systems as discussed in Section 4.5.

In some instances the set of safety tasks required for the new system cannot be achieved by augmenting tasks already listed in other program plans. In such a case, the next step is to prepare a matrix listing those subjects that need to be studied to determine which activities require development of new safety tasks. The criterion used for including a subject in the matrix is whether a process, a facility, an aspect of the design, the inherent nature of a material being used, or some other characteristic of the system is potentially hazardous. A safety task is required if the potential hazard is not already adequately controlled by available resources incorporated in the system. Having determined the set of subjects for which safety tasks need to be developed, the system safety planner must describe the tasks needed and prepare procedures for transforming the tasks listed in the matrix into areas of assured safety. The fourth level in Figure 4-5 represents this process.

4.8 SYSTEM SAFETY DOCUMENTATION

Item 5 of Appendix C is concerned with the documents that need to be issued in response to the requirements established by the system safety program plan. It should be noted that, with few exceptions, the generation of a document is not in itself completion of a safety task. Most often a document is a report of the status of an analysis, mathematical or experimental.

A system documentation tree is presented in Figure 4-6. As the figure indicates, the system safety organization, in addition to publishing material that is concerned solely with safety, may provide material for inclusion in publications prepared by other organizations. All such published material should be traceable through the system safety plan to requirements established in the contract. Among the more important safety documents released are those which describe changes to the design, manufacture, operation, or

Figure 4-6
System Safety Documentation Tree

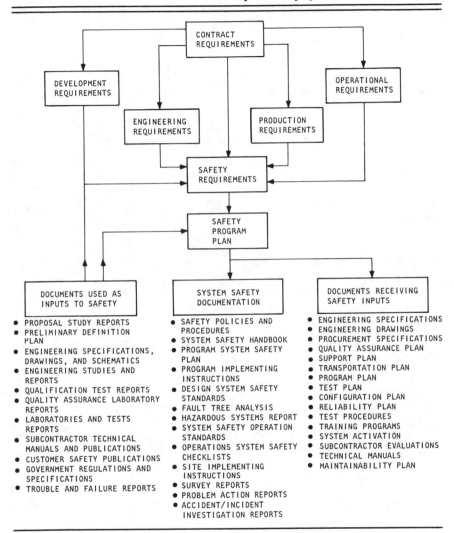

maintenance of the the system. Such documentation can be applicable to the system during early phases, before drawings are released, or during later phases, when they force changes in drawings or hardware.

4.9 SYSTEM DESCRIPTORS

The part of the program plan that is concerned with controls, item 6 of Appendix C, discusses those tasks that are carried out to

1. Assess whether the tasks performed to achieve and measure the required safety level are adequate in scope and in execution.
2. Ensure that all safety tasks are completed in a timely fashion in relation to the system life cycle, so as to achieve optimum effectiveness of each task.

Establishing a set of controls, that is, assessing adequacy and determining timeliness, requires an understanding of the system and the program for its development. There are certain techniques by which relatively small amounts of information can be used to describe large systems containing millions of parts, made of hundreds of different types of materials and processes, worked on by thousands of people, and used for many years under varying conditions, to be understood through the use of relatively small amounts of information. Basic information needed for understanding a system, regardless of the form of the system or the use to which it is put, consists of a description of

1. Structure, the interrelationships among the elements of a system as defined by parameters such as relative importance, decision-making properties, space, and time, in accordance with the manner in which each parameter is influenced by the entire system.
2. Distinguishing characteristics, qualitative and quantitative, that describe those variables needed to characterize the system uniquely and differentiate it from other systems that are similar in form or function.

4.9.1 System Structure

The structure of a system can be represented in graphical or literal form. Graphical representations of information about the structure of a system that have been found useful in system safety studies include

1. Series-parallel structure
2. Source-sink structure
3. Decision structure
4. Hierarchical structure
5. Time sequence structure
6. Logic structure
7. Information flow structure
8. Open loop–closed loop structure
9. Signal flow structure
10. Matrix structure

These structural representations lend themselves readily to computer processing. Some description of graphical representations and the manner in which they relate to system safety is presented next.

4.9.1.1 Series-Parallel Structure

Figure 4-7 shows the simplest form of the series and parallel structures. In the series relationship, Figure 4-7a, *A* precedes *B* so that the output of *A* acts as an input to *B*. The inability of either *A* or *B* to function properly degrades or entirely prevents the output event from occurring. The parallel relationship, Figure 4-7b, implies that the input event is presented to both *A* and *B* and that satisfactory operation of either *A* or *B* permits the output event to occur.

The blocks of Figure 4-7 may be used to represent hardware elements or morphological events, or the figure may be a model for some combination of elements. In the latter instance, each block may represent a major component of a system or an entire subsystem. In addition, it is permissible to mix blocks in the figures so that both equipment and morphology are represented in one figure or one block. For example, in Figure 4-7b, *A* could represent an autopilot and *B* could represent a pilot, with either capable of successfully accomplishing a required output event such as maintaining a specified heading.

A graphical representation of a system made in this fashion is called a "block diagram." Modeling a system through the use of block diagrams is of assistance in determining which elements of a system are most dominant in terms of their ability to give rise to hazardous occurrences. This notion of dominant paths leading to a hazardous occurrence is expanded further in the sections dealing with fault trees.

Figure 4-7
Series-Parallel Structure

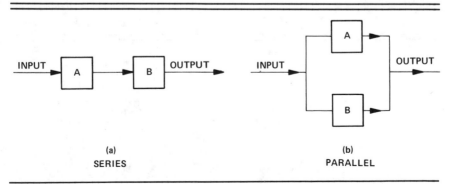

(a)
SERIES

(b)
PARALLEL

4.9.1.2 Source-Sink Structure

The graphical representation that describes "sources" and "sinks" is illustrated by Figure 4-8. In the safety domain, this type of representation is useful to describe the flow of those materials and energies which, by themselves or in combination with others, may give rise to hazardous occurrences. For example, the sink in Figure 4-8b could represent an enclosed space, such as a compartment in a battery or ship or a human lung, and paths *b* and *c* could represent paths of flow for fluids or electrical energy. If, for example, a combustible gas flows along one path and electrical energy flows along the other, it is necessary to ensure that simultaneous flow does not occur. The latter hazard is applicable to the human lung during surgery employing combustible anesthetics. Sources and sinks are also used in fault trees and are discussed in the fault tree section of this book.

4.9.1.3 Decision Structure

A graphical representation of the decision structure is shown in Figure 4-9. The usefulness of this representation is that there are a number of sequential, discrete activities that must occur during the lifetime of any system, both internal and at its interface with the outside world, for which alternative choices are available. However, the ability to select a desirable choice at a given decision point may depend upon selected previous choices. For example, suppose it has been determined a priori that decision *h* shown at level IV of Figure 4-9 is an optimal choice. The ability to select decision *h* exists if and only if decisions *a* and *c* are selected at levels II and III, respectively. Conversely, Figure 4-9 shows that decision *h* is impossible to make if decision *b* is made at level II or decision *d* is made at level III.

It can be seen that Figure 4-9 appears to be the result of combining a set of flow diagrams of the type shown in Figure 4-8. Indeed, flow and decision making can be combined into one form of graphical representation, and this

Figure 4-8
Source-Sink Structure

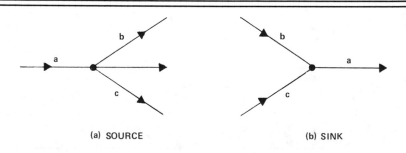

(a) SOURCE   (b) SINK

Figure 4-9
Decision Structure

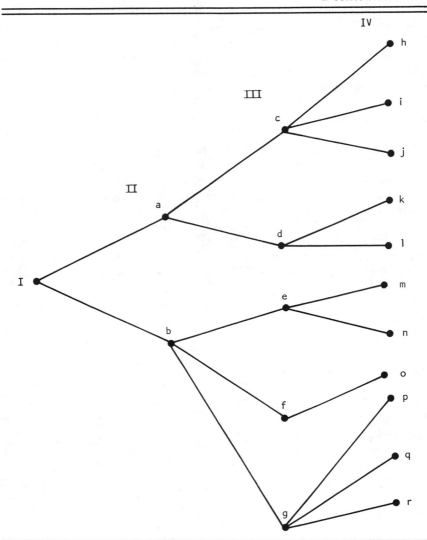

combination is used in fault tree analysis. It should be noted, however, that although the models discussed in this section are graphical in nature, the structural relationships of the system represented by the model are not necessarily graphical. Examples of system structures that are not graphical include equipment specifications, job descriptions or functions and responsibilities, constitutions and bylaws of organizations, and business policy guides.

4.9.1.4 Hierarchical Structure

The hierarchical structure is illustrated in Figure 4-10. In use, a hierarchical structure can represent entities such as organization of personnel, material, resources, money, or machines. One use for this representation is to describe organizational structure of personnel in conventional fashion to ensure that each function, task, and responsibility needed for achievement of a specified safety level is assigned to a specific individual or organizational entity.

4.9.1.5 Time Sequence Structure

Figure 4-11 illustrates an information-reporting technique that is useful in describing the time sequencing of series and parallel tasks and their inter-relationships. In this figure, task *a* precedes task *b*, which precedes task *c* in time; and similarly, task *e* precedes task *d*, which precedes task *g*, which precedes task *h*.

Figure 4-10
Hierarchial Structure

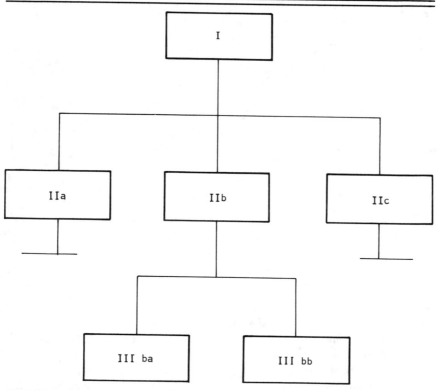

Figure 4-11
Time Sequence or PERT Network

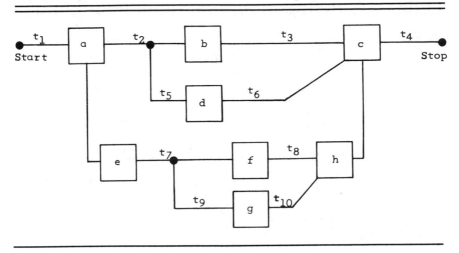

As noted in Section 4.2, a system safety program plan describes the activities required, indicates how they are to be accomplished, and establishes times when the results are required. The time sequence structure is suited for controlling the "when" content of a program plan, which can be critical in safety considerations. The purpose of safety analyses can be to establish an intrinsically safe system, design a system to minimize hazards, install safety devices or warning devices, or establish suitable emergency procedures. The ability to select the more desirable options is a function of the timeliness of the safety analyses in relation to the progress achieved in designing and constructing the system.

4.9.1.6 Logic Structure

Figure 4-12 illustrates one of the several logic representations used for describing the manner in which events that take place in a system are interrelated and controlled. In this figure, IN represents the discrete data that describes an event under consideration, and REF represents an arbitrary reference against which the input is compared. YES and NO represent the two alternative outputs whose occurrences are dependent upon the relationship between the input and the reference. This type of representation is useful in describing the circumstances under which hazardous events can occur and is further discussed in the chapters dealing with fault tree analysis.

Figure 4-12
Logic Structure

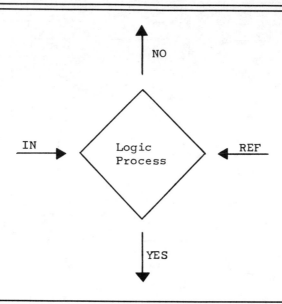

4.9.1.7 Information Flow Structure

A procedure for providing system information when the safety of the system elements is influenced by their topographical arrangement and the direction in which information flows between them is shown in Figure 4-13. In this figure flow can occur from element I to element II, but not in reverse. However, flow can occur in either direction between elements II and III. Any of the directional constraints may change during the life of a system. Information about the distance between the various elements can be expressed as shown by designations d_1 and d_2 in Figure 4-13. This method for presenting information about the system is applicable to geographical and morphological considerations. An example of the latter is the network of blood vessels in the human system, some of which have valves that can be open or closed, depending on physiological requirements.

4.9.1.8 Open Loop–Closed Loop Structure

A format for indicating whether the elements of a system are in open loop or closed loop configuration is presented in Figure 4-14. It might be noted that the blocks in Figure 4-14a appear identical to the ones in Figure 4-7a. One

Figure 4-13
Topographical Structure

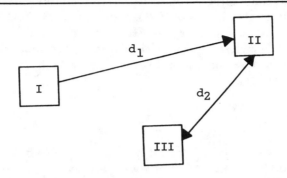

distinction between the two is that the series configuration (Figure 4-7a) describes the physical arrangement of system elements, while the open loop structure can represent a time sequence of events flowing from A to B regardless of how the system elements are configured. Similarity also exists between Figures 4-7b and 4-14b. The parallel structure in Figure 4-7b may represent the physical arrangement of the system elements. However, activi-

Figure 4-14
Open Loop–Closed Loop Structure

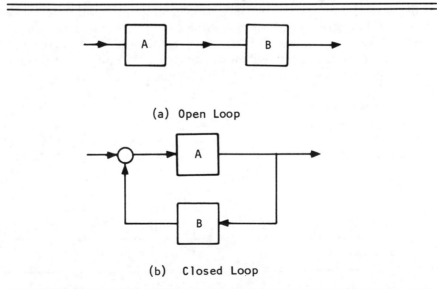

(a) Open Loop

(b) Closed Loop

ties or events represented by the blocks of the parallel structure are essentially independent of each other, and the character and timing of the outputs of the elements of a parallel structure may be different. On the other hand, in the closed-loop structure (Figure 4-14b) the events represented by block *B* exert an influence over the events represented by block *A*. Further, the overall output of a closed loop structure is derived from some combination of events where the combination takes place in accordance with the fact that block *A* is influenced by block *B*.

4.9.1.9 Signal Flow Structure

Assistance in the derivation of the block diagrams shown in Figure 4-14 can be obtained through preparation of signal flow diagrams of the type illustrated by Figure 4-15. A signal flow diagram indicates how information flows along the elements of a system and how the elements are related topologically. In the figure, information originating at points *a*, *b*, and *c*, (identifiable in the system by location or event) and the lines between these points represent activities or changes that take place among the locations or events. Therefore, in the figure *b* is a function of *a* and *c*, while *c* is a function only of *b*. The emphasis of the signal flow diagram is upon those dependencies which exist among the elements of a system rather than upon identification of the type of dependency.

4.9.1.10 Matrix Structure

A literal representation for describing system structure that is used for all systems is the matrix structure, presented in Figure 4-16. In its most elementary form a matrix structure is a method for recording data and records. The column headings A, *b*, and C and the row labels 1, 2, 3, and 4 make it possible to single out any element in the figure. For example, element 2B is located at

Figure 4-15
Signal Flow Structure

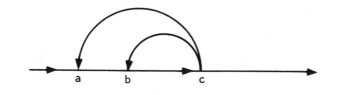

Figure 4-16
Matrix Structure

	Column A	Column B	Column C
Row 1	1A	1B	1C
2	2A	2B	2C
3	3A	3B	3C
4	4A	4B	4C

(a) RECORDS AND REPORTS

	A	B	C
1	1A	1B	1C
2	2A	2B	2C
3	3A	3B	3C
4	4A	4B	4C

(b) MATRIX

the intersection of row 2 and column B. Figure 4-16b presents a matrix for the same information contained in Figure 4-16a. However, the format of Figure 4-16a is generally used for presentation of nontechnical information, while the format of Figure 4-16b is more applicable to scientific and engineering information. One of the uses of this format in the domain of system safety is discussed in Section 4.9.3.

4.9.2 Distinguishing Characteristics of a System

The "signature" of a given system, its distinguishing characteristics, may be derived by combining a set of graphical representations with a set of value judgments and a set of qualitative and quantitative distinguishing qualities. Together, these provide a unique description of a given system. In many instances, the ability to synthesize and formulate graphical representations for a system requires an understanding of some of these distinguishing qualities. It is not practical to attempt to catalog all possible distinguishing characteristics that may be useful to a system analyst attempting to define the "signature" of a given system. However, a brief description of some qualities that are useful to the safety analyst are discussed next. Some, considered to be part of the kernel of system safety analysis, are discussed in greater detail in Chapter 9.

Perhaps the most fundamental characteristics employed to describe a system are physical, mechanical, chemical, electrical, optical, acoustical, and hydraulic. The manner in which data about physical characteristics are added to a graphical representation, so that the combination becomes uniquely related to a given system, is illustrated in Figures 4-17 and 4-18. Fig-

Figure 4-17
Sink Structure with Physical Qualities

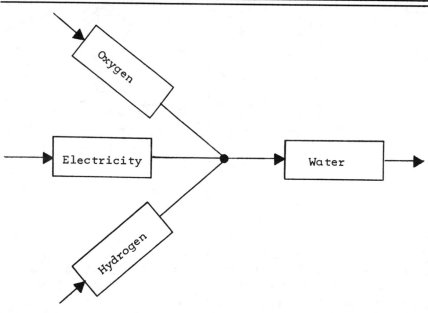

Figure 4-18
Logic Structure with Physical Qualities

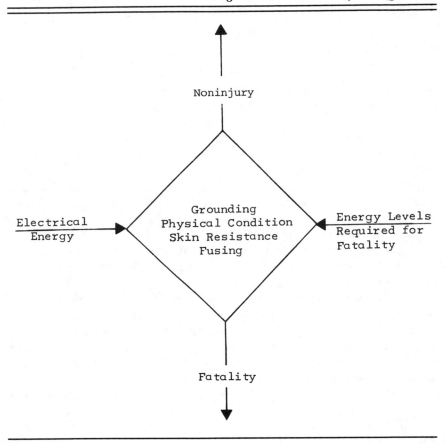

ure 4-17 presents a graphical representation of a sink structure in which the physical entities oxygen, electricity, and hydrogen and the result obtained by combining these three, water, are identified. It may also be desirable to provide additional physical characteristics, such as weight, voltage, amperage, and isotopic content of the various entities shown in this figure.

Figure 4-18 presents a logic structure of a part of the system to which certain physical quantities have been added. The YES, NO outputs of the logic structure are fatality and noninjury, respectively. The input event refers to electrical shock such as would result from either carelessness or faulty equipment. The reference relates to the amount of electrical energy necessary to cause injury or fatality to the average human. The logic conditions refer to some of the factors which operate in combination with the two inputs such that one or the other of the two outputs occurs. Here too it may be desirable to add other physical descriptors such as frequency, voltage, or current.

4.9.3 Value Judgment

Another distinguishing quality that needs to be expressed explicitly and, preferably, quantitatively to define a system relates to value judgments, such as cost value, esteem value, and exchange value, and to the optimization process by which these values are related to the safety and effectiveness of the system. Value and optimization of safety were discussed in Chapter 2.

The application of value judgment as an addition to a graphical representation may be illustrated by considering a hierarchical structure represented by an organization chart. One measure of the value assigned to each element is its vertical distance from the top element of the chart. Those systems for which control of safety activities has a high value can be expected to appear at high levels in the hierarchical structure. In some instances the safety organization is located at level II (refer to Figure 4-10).

Another relationship that influences value judgments made about a system concerns the frame of reference of the evaluator. For example, value judgments about the safety of a home appliance may vary depending upon whether the judgment is being made by the supplier of the raw materials, the manufacturer of the appliance, the distributor, or the purchaser. Differing value judgments based on reference frame are implied in the negotiations that take place between higher and lower levels of a hierarchy about the amount of resources considered necessary for carrying out the functions and responsibilities assigned to the lower levels. Differing value judgments about the safety of a given system need to be integrated into one requirement, regardless of reference frame.

4.9.4 Variations in Physical Characteristics

A set of undimensioned graphical representations which provide knowledge about the structure of a system, when combined with information about physical characteristics and value judgments, only yields information about the nominal character of the system. A more complete analysis requires additional data of the type shown in Figure 4-19 describing the range over which nominal values can vary along with the manner in which such variation is distributed throughout the range. Knowledge of this information could be critical in the safety domain even for parameters such as length or volume that are routinely accepted without question.

The probability distribution is commonly used to describe variations about some nominal value of a given parameter. Probability distributions

Figure 4-19
Range and Distribution of Nominal Value

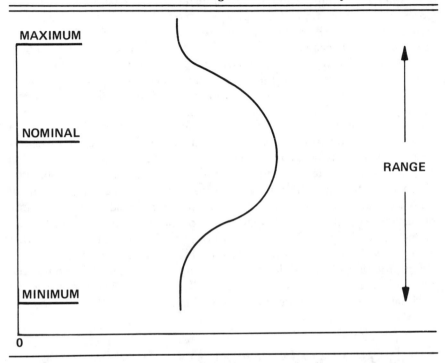

compensate for the fact that a nominal value of any parameter, even if explicitly specified, is never known with absolute precision. Rather, there is some degree of uncertainty brought by imprecision in measuring, the manufacturing process, maintenance procedures, aging, or environmental factors. In addition, the use of value judgments, because they tend to diffuse the specification of the nominal figure of merit, also introduces some uncertainty. This notion of imprecision in the nominal values of parameters used to describe the system and some commonly used probability distributions are discussed in Chapter 9.

The amount of uncertainty associated with the set of parameters used to describe a given event is increased when initiation of the event requires prior completion of another, for the second event must also have some uncertainty associated with the parameters needed for its definition. This notion of prior quality that needs to be expressed explicitly and, preferably, quantitatively in order to define a system depends on the value judgments that are made for each event. The basis for making value judgments, such as cost value, esteem value, and exchange value, and the optimization process by which these values are related to the safety and effectiveness of the system were discussed in Chapter 2. The implication of such value judgments in combination with a graphical representation of the system is considered later in Chapter 9. Intuitively it can be seen that the need for two events to occur in a specific order is likely to decrease the probability of occurrence of the combined event. Concomitantly, the precision with which the probability of occurrence of a combined event may be predicted is diminished. A definition of "conditional probability" is presented in Section 8.4, and the application of conditional probability to safety modeling is considered in Chapters 6 to 8.

One further uncertainty occurs when the parameters used to describe a system are not static or correlated as a function of time, a condition that is generally true. It is necessary in such cases to take time into consideration when describing system outputs, and this may force the use of probability distributions to define the amount of diffusion that exists about the nominal time parameter. In some cases there may be a compounding of uncertainty of occurrence of the primary event owing to a need to combine uncertainties in the nominal value of time with uncertainties in nominal values of other parameters. In such instances it may be helpful to employ the notion of "joint probability" to describe the probability of occurrence of a compound event, one which can be divided into two or more simple events. A discussion of "joint probability" is also presented in Chapter 9.

4.10 SYSTEM SAFETY CHECKLISTS

If the system is well understood and thoroughly described, it is feasible to establish a scheme for assessing the adequacy of scope and execution of the system safety effort. The precise procedure used for making such an assessment must relate uniquely to the system under consideration. The use of system safety checklists has proven helpful for this purpose. An illustration of such a checklist, proposed for use at high organizational levels, is presented in Appendix A. The elements on the checklist are presented in Table 4-1. The column headings of this table refer to four of the phases that make up a system life cycle (see Figure 4-2). The row headings describe some of the various functions and organizational entities that are required for conceiving, developing, building, and operating a system. The number pairs shown in Table 4-1, the elements of the matrix, are used for identification. The first number of each pair represents the row position and the second represents the column position of the matrix. For example, the element (3, 2) represents the third row, system safety, and the second column, contract definition phase. In essence, these numbers act as a table of contents that refer the user to the various elements in the checklist.

Table 4-1
Matrix for Assessing Adequacy of Execution

	Concept Formulation	Contract Definition	Development and Production	Operations
Program Control	1, 1	1, 2	1, 3	1, 4
Product Integrity	2, 1	2, 2	2, 3	2, 4
System Safety	3, 1	3, 2	3, 3	3, 4
Engineering	4, 1	4, 2	4, 3	4, 4
Quality Assurance	5, 1	5, 2	5, 3	5, 4
Testing	6, 1	6, 2	6, 3	6, 4
Sales	7, 1	7, 2	7, 3	7, 4
Product and Field Service	8, 1	8, 2	8, 3	8, 4

4.10.1 Expansion of Checklist Matrix

Additional detail can be provided to the matrix and, consequently, to the table of contents by expanding the rows or columns of Table 4-1. Table 4-2 is an expansion of the third row of Table 4-1, system safety, and illustrates one scheme by which such additional detail can be provided. It should be noted that one of the characteristics exhibited by this expansion is the ease with which each alphanumeric entry in Table 4-2 may be related to its origin in Table 4-1.

Illustrative checklists are provided in Appendix B for elements 1 and 3 of Table 4-1, program control and system safety, respectively. Section 1.1 of Appendix B includes a format that can aid in using the checklists. In this format the left-hand section of each box indicates the percent of the task completed and the right-hand section names the individual responsible for

Table 4-2
Expansion of Matrix for System Safety

System Safety (3)	Concept Formulation	Contract Definition	Development and Production	Operations
Management	3A, 1	3A, 2	3A, 3	3A, 4
Safety	3B, 1	3B, 2	3B, 3	3B, 4
Design	3C, 1	3C, 2	3C, 3	3C, 4
Specification	3D, 1	3D, 2	3D, 2	3D, 4
Checking	3E, 1	3E, 2	3E, 3	3E, 4
Integration	3F, 1	3F, 2	3F, 3	3F, 4
Reliability	3G, 1	3G, 2	3G, 3	3G, 4

conducting the task. Additional data, such as whether the task is ahead or behind schedule, can also be shown. The data entered on the right-hand side can be attribute rather than variable in form. For example, *yes* or *no* can be entered rather than percent complete.

The first item in the appendix, 1.1, corresponds to position (1, 1) of the matrix shown in Table 4-1, program control, concept formulation phase. Subsequent subdivisions of item 1.1 correspond to the subdivisions shown in Table 4-2. The appendix continues with suggestions for tasks to be carried out during the contract definition, development and production, and operations phases to ensure that program control activities incorporate system safety criteria. The appendix also includes suggested checklists for system safety, engineering, and quality assurance.

It is not considered necessary to repeat the form presented in Section 1.1 of Appendix B on every page of the appendix. Further, to make the appendix more general, the matrix numbering system established by Table 4-1 is omitted from the appendix after item 3.1. In addition to its use as an aid in assessing the scope of a program, a checklist can aid in assessing the adequacy of a safety program. This claim is based on the presumption that an individual making entries in the form shown in Appendix B will not consider an item to be complete if it does not appear to meet certain standards of adequacy.

4.10.2 Application to Individual Systems

An examination of Appendix B shows how emphasis is placed upon completeness of the checklist. Many of the items presented in the illustrative checklists are common to most systems and, therefore, would be included in a general-purpose checklist. However, a checklist's greatest usefulness is its development to relate uniquely to a specific phase in the life cycle of a given system, for use by a specific organization.

In using the material of Appendix B as a source document for developing checklists oriented toward specific programs it should be noted that

1. No information is provided for the items listed in Appendix B to indicate when each task should be initiated. Since a single phase of a system life cycle can require as much as several years for its completion, it is necessary that individual tasks be started in accordance with some logical schedule. In particular, it is essential that any task whose results are required for initiation of other tasks be started in sufficient time for results to be available when needed, although it should be kept in mind that initiating a task too early risks unnecessary expense. Aiding program control through sequencing and timing of a set of system safety tasks is discussed at greater length in the next section.

2. Many of the individual items presented in a given checklist are repeated in other checklists. For example, item 1.1 in the checklist shown in Appendix B is intended for use by a system safety manager during the concept formulation phase, and item 1.2.1 in this checklist is intended for use by the system safety manager during the definition phase. Both items relate to the task of reviewing contractual documents to determine what system safety requirements are established by such documentation. This task can logically be included in many of the checklists defined by the matrix shown in Table 4.1 and augmented by Table 4-2. It is preferable, however, that each checklist be a complete entity unto itself.

It can also be seen from an examination of Appendix B that as a system advances through the various phases of its life cycle, the checklists concentrate more on subsystems and components than on systems and the tasks become more strongly oriented toward industrial safety consideration.

4.11 TIMELINESS OF SYSTEM SAFETY TASKS

In addition to providing assurance that system safety tasks are adequate in scope, control over a system safety program involves assuring that all safety tasks are completed in a timely fashion. Ideally, such timing is optimum in relation to the benefits derived by carrying out the safety tasks. For example, hazards which can only be eliminated or controlled by modifying system design can be implemented at modest cost during early stages of the design effort. Implementation of such changes by retrofit, particularly if hardware is in consumer possession, can be disastrous.

There are a variety of techniques suitable for controlling the timing of program tasks. One generally used method which is particularly helpful in controlling large, complex systems is the Program Evaluation and Review Technique (PERT), or time sequence representation. In a PERT network, each activity takes on one of several possible forms. Two of the most commonly used forms are shown in Figure 4-20. In general, the rectangle displays PERT data for events whose outputs are used mainly by the organization conducting the task described by the block. The hexagon is used to display events whose completion requires an interface between two or more functions or organizations. In Figure 4-20, the rectangle is used as a legend to display the symbology and nomenclature conventionally used in a PERT representation, and the hexagon illustrates typical entries in such a representation. When preparing a PERT representation, it might be noted that

1. Not all events in a PERT chart are of equal importance. However, events which are considered to be of great significance can be separately

Figure 4-20
Elements of a PERT Network

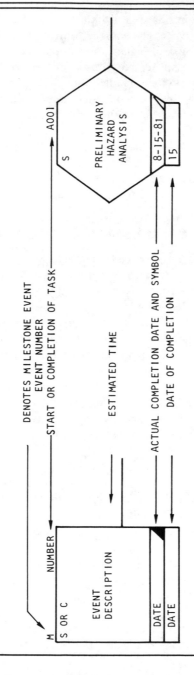

compiled to form a set of program milestones. Subsequently, such events may be identified in the PERT network by the letter M, as shown in Figure 4-9.

2. A PERT representation can provide information about the relative complexity of individual tasks and *of* the resources allocated for their completion by comparing scheduled completion dates with actual completion dates. For ease in reading the PERT charts, a diagonal cut in the "actually completed" block is filled in when the event is completed.

3. The "estimated date of completion" is not expected to be invariant. Rather, the estimates may change as additional information is gathered about the ability to complete a given task. Since an increase in time between any two events can affect the starting time of subsequent events, the relative importance of events can change when completion dates are changed.

4.11.1 Critical Path in a PERT Network

Any chain of events in a PERT network with time intervals containing zero slack time is considered a "critical path." Slack time is the difference between the length of time scheduled for an event and the length of time required for its completion. If required time is less than scheduled time, the resulting slack time is positive; if the required time is greater than the scheduled time, the slack time is negative. The critical path in a PERT network is the longest time path between the beginning and end events. This path defines the sequence of events that control the amount of time needed to complete the effort described by the PERT network.

In fact, one purpose of a PERT network is to determine this critical path so that overall progress on a program can be assessed and efforts undertaken, if needed, to ensure its completion on time. The notion of critical path in its general sense is a valuable tool in controlling the timeliness and importance of system safety tasks as well as tasks scheduled in other domains. This more general notion of critical path as it relates to the domain of system safety is expanded upon in the sections dealing with fault trees.

4.11.2 Illustration of PERT Network

The manner in which PERT events of the type shown in Figure 4-7 are combined to form a network is illustrated in Figure 4-21. In actuality, Figure 4-21 represents part of a complete PERT network, for it is truncated on the right-hand side and at the bottom, as indicated by the arrows. The left-hand side of the figure lists some of the tasks usually scheduled in a system safety program for a large system. For convenience, all subevents and subtasks associated with each task listed on the left-hand side are drawn at about the

Figure 4-21
Portion of a PERT Network

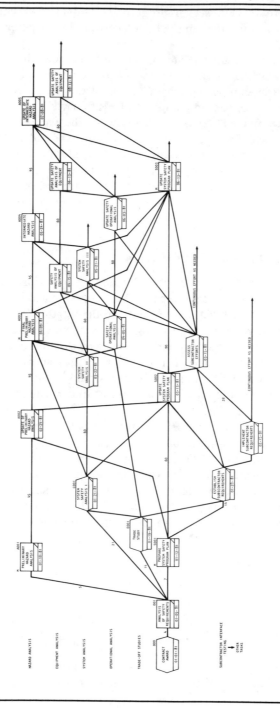

same level as the task. Consequently, the current status of any task listed on the left-hand side of the figure can be determined by examining the status of the set of subevents shown along that line.

Diagonal lines leading into an element of a PERT network indicate that initiation of the tasks defined by that element is dependent upon data or results developed through the completion of other tasks at other levels. Such other tasks may be noted on the same or separate PERT networks. Conversely, diagonal lines emanating from a given element indicate that the results obtained by completing the tasks defined by that element are required to initiate tasks defined by some other element of this or another PERT network.

It may be helpful to expand a PERT nework by a type of vertical integration carried out by combining a PERT representation with a hierarchical representation, as shown in Figure 4-22. For example, event A001, relating to preparation of a preliminary hazard analysis, can be completed by combining a subset of a preliminary hazard analysis, in which each member of the subset represents one subsystem. In such a hierarchical representation, event A001 could be considered Level I in relation to the Level II set of preliminary subsystem hazard analyses. In similar fashion, each of the events of Figure 4-21 may be represented in a hierarchical configuration as a Level I event in relation to its corresponding set of Level II events. In addition, each such hierarchical representation can be further expanded vertically to obtain whatever level of detail is needed for control of a system safety program.

It may also be seen from Figure 4-22 that several Level II PERT networks can be prepared, one for each subsystem. "Vertical integration," then, is this combination of vertical expansion through a hierarchical representation with the PERT networks established for the various subevents. Figure 4-10 employs subsystems for the representation at the second level. However, the form of a hierarchical structure used for vertical amplification can be any logical configuration which describes the relationship between the top-level events of a PERT representation and those subevents which need to be combined to produce a top-level event. If vertical integration is carried through to completion, the results will provide a time-sequenced schedule for each individual task and for the combination of all tasks. The resulting combination delineates the time sequence described by the top-level PERT network.

4.12 CHECKLIST-TIMELINESS CONSIDERATIONS

The checklist and timeliness considerations illustrated in Figures 4-21 and 4-22 can be combined into one control mechanism as shown in Figure 4-23. The elements of this figure are the following:

Figure 4-22
"Vertical Integration" of PERT Representation

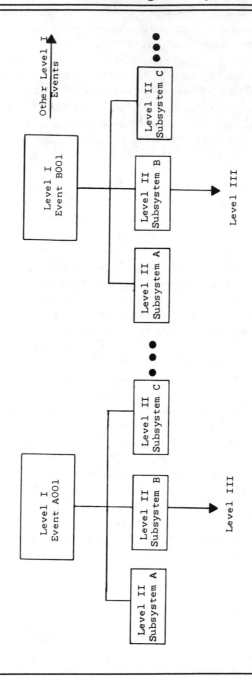

Figure 4-23
Checklist–Time Line Considerations

Figure 4-23
Checklist–Time Line Considerations

TASKS	ORGANIZATIONAL RESPONSIBILITY								TIME PHASE					
	PROGRAM MANAGEMENT	SYSTEM SAFETY	MANUFACTURING	QUALITY ASSURANCE	TESTING	ENGINEERING	CUSTOMER REPRESENTATIVE	OPERATIONS	1	2	3	4	5	6
PREPARE AND UPDATE SYSTEM SAFETY PROGRAM PLAN	A	P	S	S	S	S	S	S	X	X	X	X	X	X
SYSTEM ANALYSIS	C	S		C	S	S	S	S	X	X	X			
DEMONSTRATE OPERATIONAL READINESS		P	C		P	A	P						X	X
CONDUCT HAZARD ANALYSIS		C	C	S	S	A	S	S		X	X	X		
DEMONSTRATE MAJOR PRODUCT IMPROVEMENT	A	P	C	S	P	A	C	C			X	X	X	X
PRODUCTION ACTIVITIES		S	P	S	C	C	C	C			X	X		

1. The tasks to be performed, shown on the left-hand side of Figure 4-23, are derived from a comprehensive checklist, such as the one suggested by Appendix B.
2. Organizational responsibility, the middle section of Figure 4-23, is derived from Figure 4-1, but must be modified to be descriptive of the organizational structure relating to a specific system. The abbreviations used in this part of the figure are as follows:

   P: Prime Authority. Originates requirements, makes decisions, or carries out operation.
   A: Approval Authority. A series function, may accept or reject work or action of (P).
   R: Reviews. Requests corrective action from (P) as required.
   C: Coordinates and follows up on corrective action.
   S: Supportive Services. Provides services for (P) as requested.
3. The time phase portion of the figure, shown on the right-hand side, is derived from Figure 4-2 which describes the system life cycle. Six major phases and three subphases for the contract definition phase were chosen in this figure for illustrative purposes. This number of phases and subphases is representative of programs for large systems. The actual number of major phases and subphases used for each system is established so as to provide necessary control.

4.13 SAFETY TRAINING

Safety training and motivation, an item in Appendix A, is needed to:

1. Develop an understanding of the importance of safety to each individual, the system, and all associated personnel and equipment.
2. Assure that safety practices and procedures and emergency procedures are understood and carried out properly during all phases of the system life cycle.

   The role of system safety in relation to safety training and motivation is to

1. Establish the extent of training to be provided to all personnel associated with the system.
2. Assist in the preparation and instruction of the training material. Both preparation and presentation may sometimes be enhanced by use of specialists and consultants from outside sources and agencies.
3. Perform audits to ensure that the safety training and motivation are adequate in scope and execution.

Table 4-3
Training Time Matrix

Group	Training Time
Top management	0.5–1 hour
Middle and lower management	2–4 hours
Supervisors	4–6 hours
Shop personnel	1–2 hours
Inspectors	2–3 hours
System operators	4–12 hours
System users	0.1–0.3 hours

A typical matrix illustrating representative training times is presented in Table 4-3. The groups shown in this table do not represent all groups that may need training, and the training times represent ranges for an average system. The operators of a complex, hazardous system may require months of training before authorization for use is appropriate. The user category is intended to relate to the case where the user is not an operator, such as a passenger in an aircraft.

An extensive selection of aids is available for use in training and motivation of most of the groups shown in Table 4-3. Good sources for such material include the National Safety Council, the U.S. Department of Labor, the American National Standards Institute, the Underwriters Laboratories, the American Association of State Compensation Assurance Funds, the American Insurance Association, the American College of Surgeons, the Department of Education, and the Department of Health and Human Services. A more complete listing of government, industry, and labor organizations, along with their addresses, may be obtained through the references listed at the end of this chapter. Typical training and motivation activities for these groups include

1. Review of potential hazards in the system for users, operators, supervisors, manufacturing personnel, inspectors, and test personnel.
2. Orientation programs for new personnel and for experienced personnel undertaking new activities.
3. Use of films, lectures, and newsletters for dissemination of safety information.
4. Presentation of formal safety courses culminating in certificates of completion.
5. Cash award programs for suggestions to improve safety and for the elimination of potential hazards.

Additional Reading

Blood, J. W. PERT, A New Management Planning and Control Technique. Report T-74. New York: American Management Association, 1962.

Bode, H. W. *Network Analysis and Feedback Amplifier Design.* New York: Van Nostrand, 1945.

Brown, R. G. *Smoothing, Forecasting and Prediction of Discrete Time Series.* Englewood Cliffs, N.J.: Prentice-Hall, 1963.

Chestnut, H. Information requirements for system understanding. *IEEE Transactions on Systems Science and Cybernetics*, January 1970, pp. 3-12.

Chestnut, H. *Systems Engineering Tools.* New York: Wiley, 1965.

Croxten, F., and D. Cowden. *Applied General Statistics.* Englewood Cliffs, N.J.: Prentice-Hall, 1956.

Factory Mutual Engineering Corporation. *Handbook of Industrial Loss Prevention.* New York: McGraw-Hill, 1967.

Gardner, M. F., and J. L. Barnes. *Transients in Linear Systems.* New York: Wiley, 1953.

Gingerich, D. J., Esq. Quality and the law. *Quality*, May 1979, p. 71.

Hannaford, E. S. *Supervisor's Guide to Human Relations.* Chicago: National Safety Council, 1967.

Mason, S. J. *Feedback Theory: Some Properties of Signal Flow Graphs. Proceedings of I.R.E.*, Vol. 41. September 1953, pp. 1144-1156.

Mesarovic, M. D. *Views on General Systems Theory.* New York: Wiley, 1964.

National Safety Council. *Accident Prevention Manual for Industrial Operations.* Chicago: The Council, 1969.

Nyquist, H. Regeneration theory. Bell System *Technical Journal*, January 1932.

U.S., Department of Labor. *Guide to the Preparation of Training Materials.* Washington, D.C.: Government Printing Office.

Chapter 5

HAZARD ANALYSIS

5.1 ROLE OF THE SAFETY ENGINEER IN HAZARD ANALYSES

A hazard is defined in Section 1.3 as a potential condition which may cause injury or death to people or damage to equipment or property. The fundamental role of the safety engineer is implicit in the word *potential*. Essentially, the safety engineer's task is to prevent a potential hazard from becoming an actual hazard. When prevention is not possible, the role of the safety engineer becomes one of controlling the result and consequences of the hazard's occurrence. The functions and responsibilities of the safety engineer in fulfilling this role may be described as follows:

1. Determine what potential hazards are associated with the system over its complete life.
2. Estimate the probability that a hazard may occur and the severity of injury or damage that can result from its occurrence.
3. Determine which of the hazards can be prevented by modifying the system design or changing the procedures which govern its use.
4. Determine methods for controlling those hazards which cannot be eliminated from the system by changes in design or procedure.
5. Determine the risk associated with operating or using the system, based upon the hazards that remain inherent after completion of the system safety effort to eliminate and control hazards.
6. Determine any hazards for which it is desirable to establish a monitor and alarm system.
7. Determine any hazards for which it is desirable to establish emergency procedures for escape or treatment.

In carrying out these functions and responsibilities, the safety engineer needs to establish a level of importance for the hazards that are determined to exist in the system. Level is obtained by establishing standards for some combination of the two criteria, likelihood of occurrence and severity of injury or damage. Evidence that the combination of both criteria are necessary can be obtained by considering the two extremes of each criterion.

With regard to severity, hazards are classified in Section 2.4 into the four classes IV–I, ranging from safe to catastrophic. At one extreme, it can be presumed that even if it is desirable, it may not be feasible to prevent the occurrence of all nuisance levels, Class IV and Class III hazards. At the other end of the spectrum, there may be potentially serious hazards, inherent or external, which cannot be eliminated from the system. Gasoline used in automobiles is an example of an inherent Class I hazard: under certain conditions it is capable of causing a fatality. Nevertheless, the use of this fuel can be made acceptable by providing adequate assurance that the likelihood of its causing a Class I event is acceptably low.

We can illustrate the use of the combined criteria—likelihood and severity of consequence—with the example of estrogenic birth control pills. When the pills were introduced, elaborate procedures were pursued to ensure an acceptable safety level for the product, although some controversy remains as to whether the risk level achieved is low enough. Two pharmaceutical firms, together responsible for supplying about 15 to 20 percent of the market for birth control pills, agreed to halt manufacture in 1970. In this instance, the progestational component both manufacturers used in the pills was different from that generally used in other oral contraceptives. Government experts indicated that while the components used in all the pills mimic the same normal female sex hormones, very small differences in their chemistry could produce a major difference in biological effects. These experts indicated that the probability of occurrence of the risk inferred was very low. However, the consequences resulting from the hazard's occurrence was assessed to be of sufficient importance to warrant cessation of manufacture until further evidence could be developed.

The priorities that may be assigned to hazard elimination in terms of severity and probability of hazard occurrence are shown in Figure 5-1. High priorities are established when both the severity and the probability of occurrence are high, low priorities are established when both are low, and judgment or further analysis is required in the other two cases.

5.2 RELATION OF HAZARD ANALYSIS TO SYSTEM LIFE CYCLE

Before describing a step-by-step procedure that can be followed in carrying out a hazard analysis, certain overall considerations are introduced. The first, and perhaps most important, is the relationship of a hazard analysis to system life cycle. A discussion of system life cycle is presented in Section 4.2.1, and it is depicted schematically in Figure 4-2. In this figure, which represents the general case, a life cycle is shown to contain six segments—concept formulation, contract definition, development, production, operations, and disposal. The various segments have some degree of overlap and

Figure 5-1
Priority Levels

CONSEQUENCES OF
OCCURRENCE IN
SEVERITY

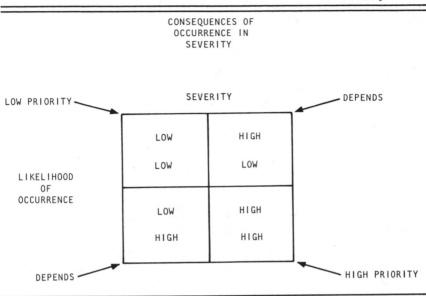

information obtained in any given segment may be fed back into segments initiated earlier.

The relationship between the system life cycle depicted in Figure 4-2 and the hazard analysis is shown in Figure 5-2. The lower portion of Figure 5-2 is the same as Figure 4-2, but the system phases and hazard analysis categories in Figure 5-2 are shown in relation to time. Therefore, the figure shows how the character of the hazard analysis changes to relate uniquely to each of the various phases of the system life cycle. The thrust of hazard analyses during early stages of the system life cycle is toward elimination of those hazards uncovered by the analysis through design modification. In subsequent stages of the life cycle, when system design becomes more detailed and production is to be initiated, the thrust of the hazard analysis tends to move toward control of hazards. This evolution of the hazard analysis occurs because hazard elimination through design becomes increasingly costlier to effect as time increases in the system life cycle.

In the initial stages of concept formulation, design changes may be effected by modification to the system drawings, and cascading effects of such changes are minimal. In later stages of the system life cycle, implementation of identical modification may require changing detail and interface drawings as well as system drawings. Consequently, cascading effects are increased, and configuration management efforts are amplified. In addition

Figure 5-2
Relationship of System Life Cycle to Hazard Analysis

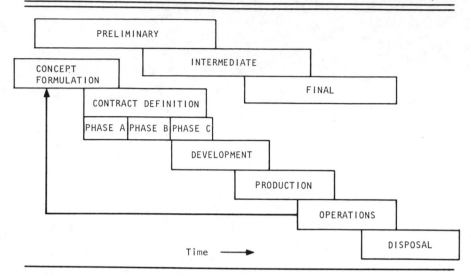

to larger costs in implementing design changes later in the system life cycle, more time is required for each change and, consequently, such changes are more likely to result in schedule slippage. In turn, schedule slippage generates its own problems which may cause unsafe qualities to be incorporated in the system.

Thus, control of hazards can be effected through design modification or establishment of procedures to control operation, maintenance, and other system activities. You can see that there could well be a motivation for hazard control through operating and maintenance procedures as the progression of time in the system life cycle increases. Hazards that are not removed from a system remain as inherent characteristics of the system and are referred to as "residual hazards." Control of residual hazards, when they are known, is often attempted through procedures designed to prevent their occurrence or through planned emergency action that prevents or alleviates the hazard's consequences should the hazard occur. It is generally not possible or economically feasible to remove all residual hazards in a system's design. By their nature, for example, conventional automobiles require combustible fuel for power. The presence of this fuel must be considered a hazard under circumstances that are known to have some probability of occurrence. Examples of activities planned to alleviate or minimize the consequences of a hazard's occurrence include firefighting procedures, descent to earth by parachute, use of snakebite kits, and certain types of surgery.

From a safety point of view, it is preferable to eliminate known hazards. Failing this, the next best course of action is to control the hazards' effects. Two approaches for accomplishing this aim are (1) minimizing probability of occurrence through design and (2) configuring the design so that if a hazard occurs, no positive action is required to ensure that its effect will not be injurious. The latter option is illustrated by installation of a relief valve on a pressure vessel. Although the hazard of overpressurization has not been eliminated, excess pressure can be vented harmlessly. A less desirable choice requires a need for taking positive action to prevent or minimize a hazard's injurious effects. From a safety point of view it is preferable that steps initiated after a hazard is identified be taken in the preceding order regardless of the point in the system life cycle at which the hazard is identified. The optimization process will dictate whether the actual course of action is hazard elimination by design, automatic control by the system, or active control by the user.

5.3 TIME-PHASED INDICATORS

In carrying out a hazard analysis within the time phases defined by the system life cycle, it may be helpful to distinguish between hazards that are leading or current in relation to the various system phases. The terms *leading* and *current* are used in the hazard analysis in a manner somewhat analogous to their use in the economic domain as an aid in predicting the state of the economy. A leading input refers to a set of hazards that can be expected to occur in a given phase but are determined prior to the initiation of that particular phase. For example, operational hazards often need to be identified as early as the concept formulation phase if they are to be avoided or adequately controlled. A current input identifies a set of hazards that can be expected to occur in a particular phase and are determined near the initiation of that phase. In some instances, a hazard can find its way into a list of current hazards because the analysis of leading hazards was not thorough enough to identify it earlier.

The economic domain also makes use of a lagging indicator to verify and supply a measure of confidence in the results obtained from the other two indicators. There is also a lagging input in the hazard analysis, but its analogy to the one in the economic domain is somewhat looser. A lagging input identifies the set of hazards whose elimination or control can be achieved only by making modifications to an existing system. That is, the hazard is determined after the system design is frozen and perhaps after the system is in the possession of the user. Sometimes lagging hazards become known because of their occurrence. The number of lagging hazards identified in a system, in combination with their integrated consequences, is in a sense a verification of the completeness with which the leading and current hazard analyses were carried out. It is, in its way, a measure of the adequacy of the safety effort conducted for the system.

Another factor to be taken into consideration in a hazard analysis is the way in which the classification of a given hazard may change as a function of time without any change in the system. For example, in 1965 a high-pressure gas transmission line failed and exploded in Natchitoches, Louisiana, killing 17 people. The pipe was being operated at 750 pounds per square inch in accordance with regulations established for Class I areas. A Class I area permits relatively low safety requirements, for it describes wastelands, deserts, rugged mountains, grazing, and farmlands having relatively low population density in a quarter-mile band on either side of the transmission line. The area in which the explosion took place was properly classified as Class I when the line was installed—21 years before the mishap occurred.

In public health hazard classification also needs to be examined as a function of time. Examples include changes in immunity that take place within microorganisms in response to repeated exposure to antibiotics and changes in immunity that take place in insects over moderate and long intervals of time in response to pesticides.

5.4 HAZARD ANALYSIS AS AN ITERATIVE PROCESS

A thorough hazard analysis has both qualitative and quantitative aspects, requires analytical and implementation types of inputs, and is completed by iteration. The iterative aspects of the hazard analysis is brought about by its need for a leading input. Hazards established during the concept formulation phase of a system for, say, the operation phase are updated, refined, and analyzed in more detail several times during the intervening phases.

The analytical effort is described by tasks 1 through 7 in Section 5.1. This part of the hazard analysis becomes more complex with each iteration for the following reasons:

1. Each successive iteration requires greater detail about hazards already described. In addition, previously unidentified hazards associated with more obscure events and more complex interface relationships must be identified as system development continues.
2. The precision needed for classifying identified hazards into one of the four possible categories increases with each iteration. During preliminary iterations, when all the needed information is not yet available, there may be a tendency to classify hazards at levels that are subsequently determined to be unnecessarily high.
3. Assessments of the probability that a given hazard will occur need to be made more accurate with successive iterations.

Initial iterations of a hazard analysis are usually qualitative in nature, intermediate iterations tend to become quantitative, and the last iterations are primarily quantitative in nature.

5.5 NOMENCLATURE USED IN HAZARD ANALYSIS

The literature discussing hazard analyses uses a variety of terms; the most common ones are shown in Table 5-1. The expressions in this table are sometimes descriptive of the type of analysis and sometimes of the time interval within which they are conducted. In many instances, the various terms presented in Table 5-1 refer to analyses that are similar in nature but whose performance is carried out by different organizations, having different interfaces with the safety of a given system. For example, failure modes and effects analyses are often conducted by reliability-oriented organizations, and crew safety analyses are often performed by organizations oriented toward human factors considerations.

In this text, the several iterations of hazard analysis are classified into one of three categories, preliminary (gross), intermediate, and final, terms that refer to time. The relationship of these iterations to the system life cycle shown in Figure 4-2 is presented in Figure 5-2. Although the terms "gross" and "preliminary" have different meanings in the dictionary, they are generally used interchangeably in safety analysis. However, the term "preliminary" is

Table 5-1
Hazard Analysis Nomenclature

Preliminary Hazard Analysis
Gross Hazard Analysis
Hazard Mode and Criticality Analysis
Failure Mode and Effects Analysis
Hazard Evaluation by Time Sequencing
Logic Diagram Analysis
System Hazard Analysis
System Hazardous Failure Mode Analysis
System Malfunction Effects Analysis
Profile Analysis
Integrated System Hazard Analysis
Crew Safety Hazard Analysis
Maintenance Hazard Analysis
Operations Hazard Analysis
Fault Isolation Analysis
Fault Tree Analysis

considered to be a more precise description of the nature of the initial iterations carried out in a hazard analysis, and it is becoming more universal in the literature. It is used exclusively in this text. A large overlap is shown between preliminary, intermediate, and final analyses in Figure 5-2, for there is very little distinction in content between the final stages of one and the beginning of the next. Administrative distinctions are sometimes made that define the three categories in terms of schedule or deliverable reports.

Hazard analyses can be conducted downward from the top or upward from the bottom. The former, known as fault tree analysis, is discussed in Chapters 6 to 8; the latter, frequently referred to as hazard mode and effects analysis (HMEA), is discussed later in this chapter. A comparison of the two types is presented in Section 6.2.

5.6 INDIVIDUAL AND COMBINED HAZARDS

In conducting a thorough hazard analysis, it is necessary to consider potentially unsafe events that can be brought about by individual hazards and by combinations of hazards. Such events will be examined in Section 6.2.2. The two of most significance are equipment malfunction and human error. The combination possible using these two possibilities can be divided into the four categories shown in Table 5-2. The most complex category, D in Table 5-2, refers to a situation in which a human error of omission or commission occurs in conjunction with equipment malfunction. The complex character of this category is due to the analytical effort required to determine which pairs of events, rather than which single events, are hazardous. On occasion, the complexity may be due to the fact that although the combination is hazardous, individually neither of the single events need be hazardous. As an example, consider how an undetected leak causing a loss in the oxygen supply of an undersea vessel, an aircraft at high altitude, or a spacecraft could adversely affect the crew's ability to function properly and cause them to commit errors that could subsequently lead to hazards.

Category D in Table 5-2 is not limited to exactly two events. Rather it

Table 5-2
Combinations of Unsafe Events

Category	Equipment Malfunction	Human Error
A	NO	NO
B	YES	NO
C	NO	YES
D	YES	YES

requires that at least one event be caused by equipment malfunction and that at least one be caused by human error. The need to consider elimination or control of hazards brought about by multiple events depends upon the likelihood of their occurrence and the severity of consequence resulting from their occurrence. In like fashion, categories B and C in Table 5-2 allow consideration of more than one event, with the same constraint in that the severity and likelihood of occurrence determine the importance of the hazard to both the system and the safety effort. No special comments are necessary about category A, which contains all single and combinatorial events not contained in the other three. Consideration of this category is necessary to ensure that systems built as designed and operated in accordance with their specifications do not nevertheless present hazards.

In order to be still more complete in conducting a hazard analysis, Table 5-2 may be modified to include "human malfunction," as shown in Table 5-3. If the analysis requires such detail, the human malfunction column of Table 5-3 can be further divided to reflect the number of persons involved. Such consideration is pertinent, for example, to the negotiations held from time to time in the railroad and airline industries in relation to the need for firefighters and third pilots as aids in achieving desired safety levels. As in the case where only equipment malfunction and human error are considered, establishing the category for the combination of the three factors given in Table 5-3 is necessary, but not sufficient, for classifying a hazard. The classification of hazards from IV through I (in accordance with Section 2.4) and the risk/impact assessment must be carried out by the safety analyst in collaboration with all concerned disciplines. The general case, in which *n* hazardous events need to be combined for a system hazard to occur, is discussed in the sections dealing with fault trees.

Because system safety needs to consider all four categories shown in Table 5-3, the hazard analysis must integrate data obtained from a variety of

Table 5-3
Combinations of Unsafe Events Expanded

Category	Equipment Malfunction	Human Error	Human Malfunction
A	NO	NO	NO
B	YES	NO	NO
C	NO	YES	NO
D	YES	YES	NO
E	NO	NO	YES
F	YES	NO	YES
G	NO	YES	YES
H	YES	YES	YES

sources. Some illustrations of these various sources are shown in Figure 5-3. In some instances, as in the case of a failure modes and effects analysis (FMEA), it may be possible to incorporate the raw data from the FMEA into a hazard analysis. In other instances, it is necessary to augment the original data with further analysis to ensure that sufficient information exists for conducting a hazard analysis. Figure 5-3 implies that interfaces between system safety and system effectiveness are maintained continuously.

A schematic representation of the outputs derived from a hazard analysis is shown in Figure 5-4. As the figure indicates, the results obtained from the various iterations are integrated, along with inputs obtained from the system time line and contingency analysis, to achieve the desired level of safety in each cycle of the system. The system time line indicates when the various elements of the system are energized or used and when the various functions performed by personnel are carried out. This is relevant to a hazard analysis since it defines those intervals of time during the system life cycle when hazards can occur. Contingency analysis describes special measures under-taken to limit the effect a hazard might have when the occurrence of the hazard is known to be a certainty. This is also needed in a hazard analysis since the actual consequences resulting from occurrence of a given hazard can be prevented or ameliorated through implementation of a contingent mode of operation.

In carrying out the iterations of a hazard analysis, it is also necessary to consider the interaction of systems whose activities merge or interface for some interval of their respective time lines. Two systems illustrate this process: a passenger airline and an airplane manufacturing facility. The life cycle of a passenger airline includes an interval in which it is manufactured (refer to Figure 5-2). The life cycle of a manufacturing facility system includes in its time line the interval needed to manufacture the passenger airline system. The hazard analysis conducted for this portion of the life cycle for each of these systems may have aspects that are complementary to each other and aspects that conflict. The occurrence of conflicting aspects may be traceable to differences in the system effectiveness goal established for each. An analogous example in the biological sciences is represented by the two systems fetus and mother.

5.7 INPUTS AND OUTPUTS OF HAZARD ANALYSIS

Consider the inputs and outputs of a hazard analysis in relation to time-line considerations. Figure 5-5 shows the flow of hazard analysis outputs as a function of time for a specific hazard. At one extreme, $t = 0$, are those tasks whose completion implies prevention of the hazardous occurrence. The ideal case is represented by a system design that is intrinsically safe. This implies equipment and procedures that are incapable of generating or releasing

Figure 5-3
Data Sources for Integration into Hazard Analysis

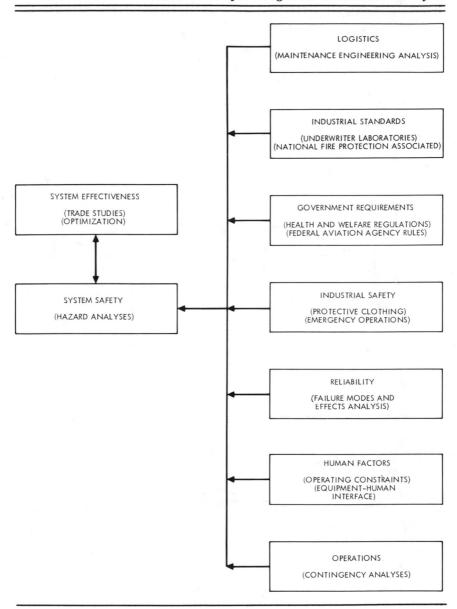

Figure 5-4
Outputs Derived from a Hazard Analysis

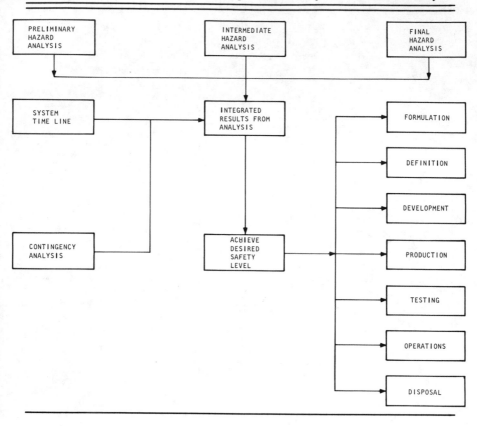

Figure 5-5
Output Time Line of Hazard Analysis

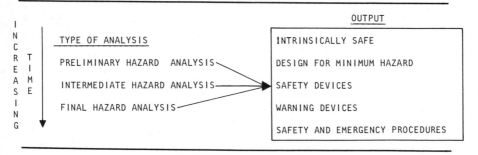

sufficient energy under normal or abnormal conditions to cause a hazardous occurrence, given equipment and personnel in their most vulnerable condition. Applying this notion of intrinsic safety to an example of electrical equipment that is required to operate in an atmosphere containing flammable gases or vapors mixed with air in quantities sufficient to produce ignitable or explosive mixtures suggests the following definition:

> *Intrinsically safe* equipment and wiring is incapable of releasing sufficient electrical or thermal energy, under normal or abnormal conditions, to cause ignition of a hazardous mixture in its most easily ignited concentration.

In this instance abnormal conditions include accidental damage to any part of the equipment, wiring, or insulation, failure or degradation of components, application of overvoltage, and adjustment and maintenance activities. Determination that a system is intrinsically safe is dependent upon knowledge of all abnormal conditions that can occur. Acquiring such knowledge about a system is, of course, the essence of the hazard analysis.

Hazard elimination or control through design that minimizes the possibility of occurrence is obviously more desirable than establishing procedures for dealing with its occurrence. However, by relating the various hazard analyses to the system life cycle, as shown in Figure 5-2, it can be seen that as time progresses, the more desirable solutions—elimination or control—become more difficult to accomplish. After a certain amount of time has elapsed, modification to the system design can only be accomplished by retrofitting. It is recommended, therefore, that system safety resources be allocated for conducting hazard analysis as early in the system life cycle as possible.

Figure 5-6 illustrates a typical time line that provides inputs to a hazard analysis. In this instance, the figure represents the operational phase of a transportation system for which n hazards have been identified. Block 5 shows the time span considered, boarding to debarkation, for, say, one trip. Block 1 indicates when during the time span of interest the ith hazard can occur. An identical time-line analysis is carried out for the remaining $n - 1$ hazards that have been determined to be possible. The time-line analysis needs to consider both recurring and nonrecurring hazards. The former is concerned with hazardous events that occur periodically during the time span of interest, and the latter is concerned with hazards that occur once or aperiodically.

Block 2 describes the activities scheduled for the various personnel operating or associated with the system. In Figure 5-6 this block shows passengers and crew. The analysis must consider all interfacing personnel—people who live near rail lines, trucking routes, and airports; those who repair and maintain the system; and pedestrians, drivers, and passengers in

Figure 5-6
Input Time Line of Hazard Analysis

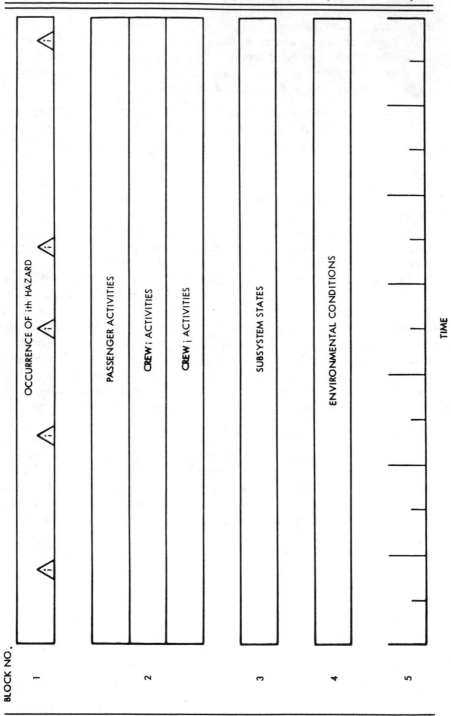

other cars, for example. The states of the various subsystems and elements are presented in block 3. The purpose of this information is to indicate when equipment is energized or de-energized, when it is off-line or operational, and so forth. This block may be subdivided to describe each subsystem and element in the system. A corresponding number of lines is then established for block 4 to describe the internal and external environments of each of the elements and subsystems described in block 3.

5.8 HAZARD MODE AND EFFECTS ANALYSIS

A hazard mode and effects analysis (HMEA), often called a fault hazard analysis (FHA) is an inductive procedure, carried out by initiating and replicating a form of the type shown schematically in Figure 5.7. Other forms are used for this purpose but the information is basically the same in each. The top of the form shown in Figure 5-7 is used to identify the system, to list the subsystem or procedures under consideration, to indicate the location of the item being analyzed, and to identify the individuals responsible for performing the analysis and implementing corrective action deemed

Figure 5-7
Format for Hazard Mode and Effects Analysis

SYSTEM _____   DATE _____

SUBSYSTEM (PROCEDURE) _____   SYSTEM RESPONSIBILITY _____

AREA/FACILITY/LOCATION _____   SAFETY RESPONSIBILITY _____

Columns: 1 ELEMENT (PROCEDURE) IDENTIFICATION, 2 FUNCTION (PROCEDURE), 3 PROBLEM, 4 CATEGORY, 5 NATURE OF INJURY/DAMAGE, 6 PERSONNEL AFFECTED, 7 CONSEQUENCE, PRIMARY, 8 CONSEQUENCE, SECONDARY, 9 HAZARD CLASSIFICATION, 10 CAUSE OF HAZARD, 11 ACTION TAKEN, 12* PROBABILITY OF OCCURRENCE

* MAY BE LEFT OUT OF INITIAL VERSION

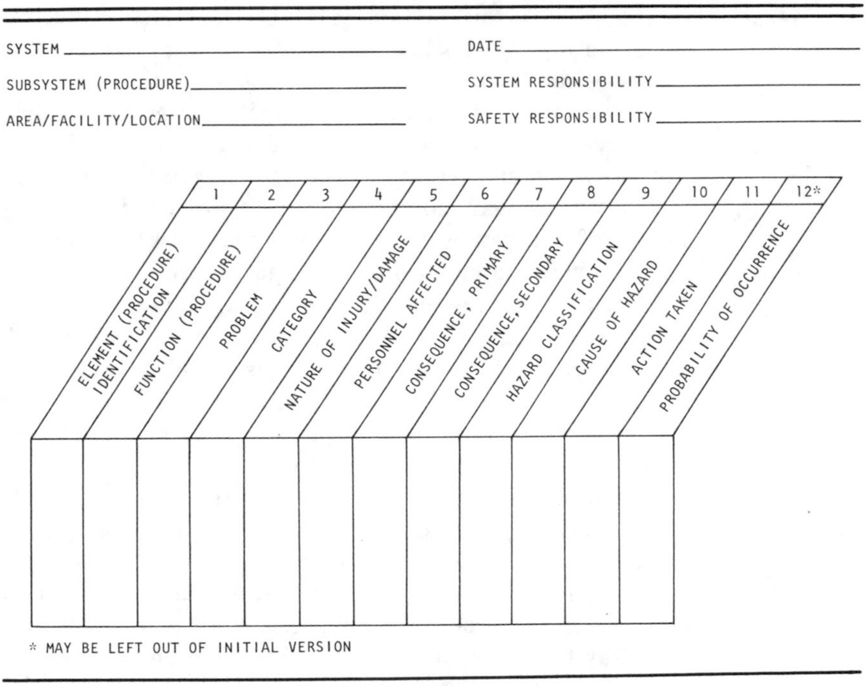

necessary. The HMEA is a bottom-to-top analysis because it is initiated by compling a list of elements in each subsystem (or procedures required for its use). In some instances a complete description of the system is not available during early stages of the HMEA.

It may not be possible to fill out the form in the order presented. Hazard classifications, for example, may not be definable until the hazard is described in some detail. In addition, it is not always possible to complete all columns of the form during preliminary stages of the hazard analysis. Asterisks are used in this figure to identify those columns often completed during intermediate and final stages of the hazard analysis. Probability of occurrence may never be determined if the HMEA is qualitative. The data required for the columns of Figure 5-7 follow:

Column 1.   The name, drawing or part number, and other pertinent identification of the element under analysis is provided here if hardware is involved. Otherwise, this column describes a procedure for subsystems which have a man-machine interface. It may be convenient to subdivide this column into two parts, one dealing with equipment functions and the other with procedural activities.

Column 2.   This column contains a description of the element's function or the purpose of the procedure. In final iterations of the hazard analysis, this column may also contain time-line information indicating when the equipment is in use or when the procedure is being carried out. Again, in the event that the human-machine interface is sufficiently large or complex, the material in this column is subdivided accordingly. This is the column that determines whether the HMEA is being conducted at subsystem, component, or detail-part level. Sometimes the first iteration is carried out at the subsystem level and additional levels are added in subsequent iterations.

Column 3.   Possible problems are hypothesized in this column. Consider, for example, a resistor in an analysis at the element level. It could be presumed to fail open or closed and to drift out of tolerance to a level that is too high or too low. At a subsystem level, one or more subsystem functions could be presumed to be absent or out of tolerance. Subsequent analysis at the subsystem level could consider whether the subsystem problems hypothesized are due to causes outside the system or to problems in subsystem components.

Column 4.   The hazard category—A, B, C, or D, in accordance with Table 5.2—is identified here. This indicates whether the hazard is caused by human error, equipment malfunction, or both.

Column 5.   A description of the injury or damage incurred is provided here. In the case of an injury, it is classified as transient or residual. In general, the effects of a problem are traced upward through the system one level at a time. A resistor failing open, for example, could cause a monitoring subsystem to fail and, in turn, cause an alarm system to become inoperative. In some instances, the HMEA form is expanded at column 5 to show the cumulative effect of a hypothesized failure as follows:

Column 5a.   Effect on next higher assembly

Column 5b.   Effect on subsystem

Column 5c.   Effect on system

Column 6.   This column describes the personnel affected by occurrence of hypothesized hazards. These could be users of the system, such as the crew or passengers of a transportation system; personnel associated with but not users of the system, such as the personnel that board passengers and maintain the system; or they could be completely unrelated people injured by debris or hazardous cargo because they happen to be in the vicinity. Satisfactory completion of this column generally requires iterations because it is necessary to know:

a.   When during the life cycle of the system its elements are tested, checked, and used and what personnel are involved in carrying out these activities. All of this information may not be available during the concept formulation and contract definition phases.

b.   What kinds of degradation could occur to the system elements as a consequence of testing, checkout, and initial use. Such degradation could create problems within the system that are not noticed until a hazard actually occurs. Consequently, this information may not be available during early iterations of the hazard analysis.

Column 7.   If not included in column 5, the hazard's effect upon the system rather than upon personnel is described in this column. This does not prevent application of the notion that the system could be wholly composed of personnel, such as in a system for mass inoculation. From the system safety point of view the following are the primary ways in which an individual hazard can affect the system:

a.   A contingency mode of operation is initiated. Depending upon the system makeup, such a mode of operation may or may not be degraded (less safe) in relation to nominal system operation.

b.   An abort is forced. The concept of abort is used here in its most general sense, for example in an automobile if an interruption occurs in a mission because a tire must be changed.

c. Abort is required but is precluded. Such a condition need not imply fatality or system loss, depending upon the nature of the system and its mission. Failure of one engine in a multi-engine aircraft after the aircraft has passed the midpoint of an ocean crossing is such a case, for the craft would then continue to its destination. The ability to initiate an abort, presuming that the system is configured to permit such an event, may depend upon the amount of warning time available. This notion is discussed in Section 5.12.

Column 8.   Any cascading effects caused by occurrence of the hazard are noted in this column. For example, a failure in voice communication between an aircraft and the ground could either require that a contingency mode of operation be adopted or force an abort. In this case the primary consequence of the malfunction, column 7, does not present a hazard to crew or passengers. However, such a problem could make the system more vulnerable in the event of occurrence of a second malfunction. Mission rules may require abort if occurrence of one problem eliminates redundant safety controls in the system. This column identifies such increases in likelihood of hazardous occurrences or the creation of new hazards that would not have existed without the occurrence of the primary hazard. As indicated in Figure 5-7, this portion of the hazard analysis may be left out of the first few iterations. The following guidelines are suggested for determining whether a biomedical problem is classified as primary or secondary; that is, whether the problem is recorded in column 7 or 8 of Figure 5-7.

a. Inability to continue with nominal or contingency events as a consequence of a biomedical problem is a primary consequence. Occurrence of a primary problem can cause injury or damage or cause initiation of contingency or emergency procedures.

b. The generation of a biomedical problem as a result of occurrence of a category A through D event (Table 5-2) is noted as a secondary consequence.

Column 9.   This column specifies the classification of the hazard, from I through IV, as defined in Section 2.4.

Column 10. The fundamental cause of the hazard is described in this column. General descriptions, such as circuit failure or electromagnetic interference, are acceptable in early drafts of the analysis. However, since action needed to prevent the hazard's occurrence requires a precise description of the cause, subsequent iterations must provide enough detail in this area to permit such actions to be taken.

Column 11. Corrective action, described in this column, will vary in accordance with factors such as the severity of injury or damage, the probability of occurrence, and the effort required to take preventive action. The action actually taken can vary from minimal, such as posting warning placards, to extensive redesign or modification in operations.

Column 12. Since a preliminary hazard analysis is by definition qualitative, it is sufficient to identify a probability of occurrence as high, medium, or low, indicating, respectively, that:

a. Corrective action must be taken to control the hazard.

b. More data are needed to determine whether the hazard needs to be controlled.

c. No action is required.

When a hazard defined by *b* is discovered early in the system life cycle, it is sometimes simpler and less expensive to change the design to avoid the hazard that it is to conduct further analysis to determine more precisely the extent of its seriousness.

5.9 CONDUCTING HAZARD ANALYSIS

Conducting a hazard analysis requires collaboration and coordination among the various organizational entities that are involved in all phases of the system life cycle. Figure 5-8 presents a generalized model, showing how various organizations relate to a hazard analysis and how the data obtained for this purpose are processed. The organizational entities shown on the left-hand side of the figure refer to associates, government agencies, subcontractors, and vendors, as well as to organizations with prime responsibility for carrying out the hazard analysis. The term "customer" in Figure 5-8 can refer to the purchaser of the entire system, the purchaser of one or more of the elements of the system, or to the user of the system. It should be noted that the purchaser and the ultimate user are often not the same for both government- and consumer-purchased systems.

The notion was introduced in Section 4.12 that the interface between the organization responsible for safety of the system and the customer varies depending upon whether the customer is also the user. In some instances, such as in the case of retail purchases, the ultimate user may have little ability to communicate directly with the organization responsible for system safety. Automobile agencies fall in the category of sales outlets that are

Figure 5-8
Procedure for Hazard Analysis

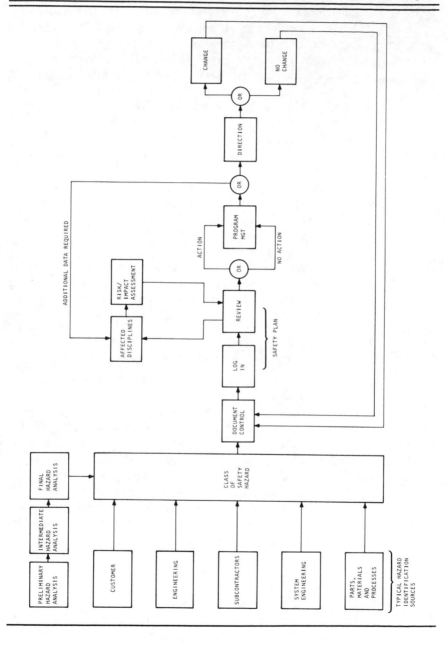

considered by the manufacturer to be the purchaser. That is, delivery of new automobiles to an agency corresponds to sale of the product by the manufacturer—there often is no mechanism for return of unsold products. Products known to be defective are recalled for modification by the dealer. Neither the dealer nor the user has significant influence on the manufacturer's safety efforts, except perhaps as a source of data based upon troubles and hazards noted by users.

In other instances, such as in the case of airplanes or cargo airliners, the ultimate user is represented by governmental agencies and in some cases by the purchaser of the system in an interface with the responsible system safety organization for hazard elimination and control. The grounding of all DC-10 aircraft by the FAA in June 1979 illustrates such representation by a governmental agency. It could be argued in the case of passenger airplanes that the purchaser of the aircraft, the airline company, is the user, but this does not diminish the character of the passenger as the primary user of the system. In the purchase of systems for government use, such as transportation vehicles, laboratory equipment, or weapons, a project office may be established for technical and fiscal control that includes representatives from user organizations. Such project offices represent the users in hazard analysis and in the case of large systems may include system safety specialists.

Integration of the data required for a hazard analysis needs to be horizontal and vertical. The horizontal axis shown at the top left-hand side of Figure 5-8 is the system life cycle time line described in Figure 5-2. The vertical axis shown directly below it refers to the hierarchy of organizations involved in the analysis, beginning with the customer, the system manager at the top, continuing through the major elements that interface with safety, and extending to basic elements at the bottom. These basic elements involve the purchase or manufacture of detail parts and raw materials and the specification or performance of the processes that make up the system. The block in Figure 5-8 with this label may or may not be the name of an organizational entity. Although not explicitly shown on the figure, it is presumed that data developed by each of the system safety organizations conducting a part of the hazard analysis are transmitted to all organizations shown or implied on the left-hand side of Figure 5-8 as well as to a central document-control point.

A procedure for conducting a hazard analysis is discussed next. The discussion is focused initially upon the operations (mission) phase of the system and is then extended to other phases of the system life cycle. Emphasis is placed on the operations phase because it is sometimes the only portion of the system life cycle to which system safety principles are fully applied. In general, this is because hazards encountered during the concept, development, and production phases of a system life cycle are often fewer and less severe than those occurring during the operations phase. Hazards associated with production and testing are often the concern of the industrial engineer and may be controlled by OSHA regulations. Nonetheless, the principles described for conducting hazard analyses are equally applicable throughout all phases of the system life cycle.

5.9.1 Hazard Compilation

The first and perhaps the most fundamental task to be performed in carrying out a hazard analysis is to compile a list of hazards that relate to the system being considered. This listing should be developed in collaboration with all organizations described or implied as "typical hazard identification sources" at the left-hand side of Figure 5-8. Such a listing, intended as an illustration of a mission or operations segment of a typical manned space flight, is presented in Table 5-4. With the exception of the last item, however, the table has general usefulness for identifying hazards. All items in Table 5-4 are considered self-explanatory with the possible exception of item 13, stress reversal, which refers to a circumstance in which a given stress is suddenly shifted from one direction to another. Examples of stress reversal include sudden decompression such as that which occurs when a driver returns to the surface rapidly, when the skin of a pressurized aircraft is punctured at high altitude, or during childbirth, which requires that the series of quiescent subsystems become functional within a short interval of time.

5.9.2 Hazard Interface with Time Line

The next step in the analysis is to relate the effects of a hazard's occurrence to the system time line. The procedure described earlier, using the format shown in Figure 5-6, may not always provide complete information. That is, an occasional unique hazard can give rise to varying results depending upon when it occurs in the time line. Table 5-5 illustrates some of the possible effects resulting from hazard 9 of Table 5-4, pressure extremes. As Table 5-5

Table 5-4
Typical Hazards for Manned Space Flight

1. Acceleration	10. Radiation
2. Atmosphere (excesses or inadequacies)	11. Shock and impact
3. Biomedical problems	12. Stress concentration
4. Corrosion	13. Stress reversal
5. Electrical shock	14. Temperature extremes
6. Explosion	15. Toxicity
7. Fire	16. Vibration
8. Noise	17. Weather
9. Pressure extremes	18. Weightlessness

shows, the effects resulting from pressure extremes vary from Class IV through Class I for both personnel injury and equipment damage. The first column of Table 5-5 shows some systems that can be affected by pressure extremes. Possible causes, shown in the center section, remain to be related to the system time line. Loss of cabin pressure, for example, a cause associated with rapid pressure change, can occur only when there exists a pressure differential between the cabin and its outside environment and can be a hazard only if the outside environment has too low a partial pressure of oxygen.

5.9.3 Intermediate Step in Hazard Classification

A method is now proposed that permits hazards to be classified more precisely in relation to mission events, allowing preliminary and intermediate hazard analyses to flow directly into the final analysis, HMEA, or fault tree. In this intermediate step the theoretical effects that each hazard can have are noted. These theoretical effects are then related to the system time line to determine what events are actually possible. As an illustration of this method, consider the fact that each of the hazards can, in theory, cause injury. Using injury as an effect, the mission events in which crew injury can occur are taken from the time line and are presented in Table 5-6.

5.9.4 Effect-Event-Hazard Matrix

The next step in conducting a hazard analysis is to prepare an effect-event-hazard matrix. To do this it is necessary to first list each undesirable effect that can occur during each mission event. An effect-event-hazard matrix is then prepared for each combination of undesirable effect and mission event that, in conjunction with the hazards previously listed (as in Table 5-4), can bring about the undesirable effect under consideration. Table 5-7 is such a matrix for one of the mission events of the manned space flight example. The undesirable effect is injury, the event is extravehicular activity, and the hazards selected from Table 5-4 are 2, 3, 9, 10, 14, 15, and 18. These matrices can usually be generated by computer and combined by computer to prepare time-line analyses.

5.9.5. Extension to Nonmission Events

In a complete hazard analysis, all events occurring during the system life cycle, which includes the subset of mission events, are listed and an appropriate compliation of hazards prepared. In the case of manned space flight, the compilation of hazards presented in Table 5-4 would be augmented to include those hazards that are unique to the nonoperations phases shown in

Table 5-5
Effects Resulting from Pressure Extremes

Hazard/System	Possible Cause	Possible Effects
Low Pressure		
Vacuum systems	Compressor failure	Unbalanced forces
High altitude vehicles	Increase of altitude with-	Collapse of pressure
Space vehicles	out pressure relief	vessels
Human cardiovascular	Inadequate design against	Inadequate breathing air
system	collapsing forces	Leaks
Circulatory system	Increase in altitude with-	Bursting of pressurized
	out suitable respiratory	vessels
	equipment	Physiological damage
	Rapid condensation of	(atelectasis)
	vapor in a closed system	Death
	Decrease in gas volume by	
	combustion	
	Cooling of hot gas in a	
	closed system	
Rapid Pressure Change		
High-altitude vehicles	Rapid expansion of gas	Joule-Thomas cooling
Space vehicles	High gas compression	Compressive heating
Underwater vehicles	Rapid changes of altitude	Explosive decompression
Compressing or pumping	Loss of cabin pressuriza-	Physiological (aeroem-
equipment	tion at high altitudes, in	bolism, cramps,
Airfoils	space or underwater	bends, chokes, eardrum
Carburetors		puncture, death)
High Pressure		
Hydraulic systems	Clogged blood vessels	Container rupture
Pneumatic systems	Overpressurization	Blast
Cryogenic systems	No pressure relief or vent	Fragmentation
Pressurized containers	Faulty pressure or relief	Propelling of container
Boilers	valve	Hose whipping
Underwater vehicles	Heating of liquids with	Blowing other objects
Engine cylinders	high vapor pressures in	(eye hazards)
Human circulatory	an enclosed system	Increased reaction rate
system	Warming cryogenic liquids	Skin penetration
	in a closed or inade-	Lung damage
	quately vented system	Cutting effects
	Impact	Shock
	Blast	Leaks in lines and equip-

Table 5-5
(Continued)

Hazard/System	Possible Cause	Possible Effects
	Container hit by fragments	ment designed for lower
	Failure or improper release	pressures
	of connectors	Dimensional changes
	Inadequate restraining	Actuation forces and
	devices	opposition
	Rapid submersion	Death
	Deep submersion	
	Water hammer (hydraulic	
	shock)	

Figure 5-2. Table 5-8, continuing with the manned spacecraft illustration, lists some nonmission events that are potentially hazardous. Some events listed in this figure can affect the crew, while others do not. Many are common to most ground-based systems. The effect-event-hazard matrices established to relate hazards and events can be expanded to describe the type of personnel affected—crew, technicians, test personnel, and so forth.

Table 5-6
Mission Events in Which Crew Injury Can Occur

1. Crew egress/ingress
2. Ground to stage power transfer
3. Launch escape
4. Maximum Q
5. Firing & separation stages
6. Ground control communication transfer
7. Rendezvous & docking
8. Ground control of crew
9. Ground data communication to crew
10. Extravehicular activity
11. In-flight tests by crew
12. In-flight emergencies involving loss of communications, power, control, or life support system; fire, toxicity, and explosion
13. Reentry
14. Parachute deployment and descent
15. Crew recovery
16. Vehicle securing and recovery
17. Vehicle inerting and decontamination

Table 5-7
Effect-Event-Hazard Matrix

Undesirable Effect	Mission Event	Applicable Hazards
Injury	Extravehicular activity	Atmosphere (2) Biomedical problems (3) Pressure (9) Radiation (10) Temperature (14) Toxicity (15) Weightlessness (18)

5.10 HAZARD CLASSIFICATION

Thus far, the discussion dealing with hazard analyses of the type described by Section 5-9 has been concerned primarily with hazard identification. The next step in the analysis is to determine the classification of each potential hazard identified. Classification is based upon two considerations:

1. The effect upon personnel and equipment resulting from the hazard's occurrence. This is defined by the classifications IV through I in Section 2.4.
2. The warning time associated with the hazard. Warning time is defined as the interval of time between identification of a problem and the occurrence of injury or damage.

Table 5-8
Nonmission Events with Potential for Injury

1. Test chamber operations
2. Assembly of major systems
3. Proof test of major components, subsystems, or systems
4. Propellant loading, transfer, or handling
5. High-energy pressurization, hydrostatic testing, or pneumostatic testing
6. Ordnance installation, checkout, or test
7. Tank or confined space entry
8. Transport and handling of end item
9. Manned vehicle tests
10. Integrated test operations
11. Static firing
12. Crew training
13. Zero gravity operations in aircraft

In a situation in which a hazard instantly results in an irreversible injury, fatality, or equipment loss, only the hazard's effect is needed for classification purposes. At the other extreme are hazards that are innocuous for some interval of time but that can eventually have Class I consequences unless adequate warning time is provided. As an illustration, consider a closed environment in which increases of carbon dioxide or carbon monoxide above allowable limits or in which the partial pressure of available oxygen decreases below allowable limits. This may occur, for example, in an undersea vessel, in scuba diving, in an airplane, and in administration of oxygen for medical reasons. In each of these examples, the hazard classification depends upon warning time, as defined above, and, if it exists, upon the sensitivity and reliability of the warning system. If little data are known during initial iterations of a hazard analysis about the capability of the monitor and alarm system to detect and provide warning, or if there is no such system, then the hazard is usually noted as Class I. Changes to higher, less hazardous levels are made, if appropriate, in later iterations when more data become available.

To pursue the circumstance in which a monitor and alarm device is part of a system, it should be noted that, strictly speaking, failure of a monitor and alarm by itself does not usually cause a hazard. Even if failure of the warning system results in a false indication and a subsequent injury occurs during emergency action undertaken in response to the false indication, it can be argued that the actual cause of the injury was procedures rather than equipment failure. In the case of carbon monoxide in a closed atmosphere, failure of a monitor and alarm system to provide warning and failure of the life support system must both occur before an injury results.

In classifying hazards according to severity and as a function of warning time, it should be noted that the amount of warning time needed to take preventive or corrective measures may not remain constant for any given hazard. It is reasonable to presume, for example, that the ability of an operator to respond to an emergency may become lessened as a mission progresses in time. In some instances fatigue, which may be expected to increase exponentially as a function of time, may be responsible for increasing the warning time needed to respond to some specific circumstance. The subject of hazard classification in relation to warning time is discussed further in Section 5.12.

5.11 HAZARD PROBABILITY

Two criteria were presented in Section 5.10 for determining the classification of identified hazards. It may also be desirable to obtain a quantitative estimate of the probability of the hazard's occurrence, especially for any Class I hazard uncovered by analysis. If two or more failures must occur before the hazard can occur, as in the illustration of excessive carbon monoxide in combination with a warning system, it may be adequate to establish boundaries and limits rather than calculating precise probabilities. In such a circumstance it is a combination of events, a joint probability, that is of interest. In general, system safety procedures require that a Class I hazard not be "closed out" without some qualitative analysis of probability of occurrence. The ability to obtain a likelihood of occurrence may be easier by fault tree analysis. (see Chapter 7). If, as a result of qualitative considerations, it is determined that a given Class I hazard is acceptably low, it does not follow that further consideration is unnecessary. The results of such consideration, presumably pursued by safety and other affected disciplines, will result in one of two findings:

1. The system has deficiencies. In this event, agreements will be reached between safety and other appropriate organizations to undertake the design and testing necessary to ensure that the deficiencies noted will be remedied. This process is represented in Figure 5.8 by the central row of boxes, between "document control" and "program management" and the feedback loop coming from the "change" box.
2. The system is adequate. This is represented in Figure 5-8 by the feedback loop coming from the "no change" box.

Success in adequately controlling a Class I hazard but needing, nevertheless, to consider its potential to act as a Class III or II hazard is illustrated by the condition of excess carbon monoxide with a monitor and alarm device in the system. Two conditions must be met to achieve adequate control of the Class I hazard:

1. The monitor and alarm system has adequate sensitivity to preclude a Class I hazard's going unnoticed.
2. The likelihood that the monitor and alarm mechanism will fail in a mode such that warning will not be given is sufficiently low. The problem of false alarm caused by defects in the monitor and alarm system is considered separately.

Suppose in the example of excess carbon monoxide that its potential effects as a Class I hazard are considered to be adequately controlled. The need to establish independent or ancillary controls for potential Class II effects for this hazard will depend upon matters such as the sensitivity of the equipment monitoring the Class I hazard, the length of time required to complete the operations segment of the system time line, and the atmosphere's constituents and partial pressures. If the Class I potential of this hazard is considered to be adequately controlled, there remains the possibility that this hazard will occur at level II or III and, becoming more serious, cascade into Class I. For example, an unnoticed increase in partial pressure of carbon monoxide could, without being hazardous in itself, induce personnel to commit minor errors that subsequently give rise to more serious hazards.

5.12 HAZARD CLASSIFICATION AND WARNING TIME

One possible entry in column 10 of Figure 5-7, "cause of hazard," is "inadequate warning time." As an example, consider an orbiting spacecraft in which the monitoring system, either in the spacecraft or on earth, has indicated an impending solar flare with radiation output of Class I intensity. If the amount of warning time is adequate, reentry can be initiated and carried out in a controlled and nominal manner. With less warning time, emergency procedures may be required, with all their attendant risks, and with no warning time, the hazard becomes a real event, with all its attendant consequences. Numerous but less esoteric examples can be obtained in the fields of medical, automotive, marine, and industrial safety.

In a circumstance in which a given hazard can be avoided provided there is enough warning time, the corrective action taken to increase the warning time needs to be adequate in relation to (1) the time needed for the "worst case" condition that can occur under nominal circumstances and (2) the time needed under those adverse circumstances that the hazard analysis indicates are likely to occur. An analysis of the type shown in Figure 5.9 is useful for describing hazards sensitive to variations in warning time. In this figure, the columns $j_1, j_2, \ldots$ represent the warning times that the system

Figure 5-9
Warning Time Requirements

WARNING TIME	MISSION TIME LINE INTERVALS			
	j_1	j_2	j_3	
i_1	E_{11}	E_{21}	⟶	
i_2	E_{12}	E_{22}		
i_3	E_{13}			

provides under nominal conditions. These times are divided into increments of time depending upon the nature of the system and the segment of its life cycle under consideration. For example, the warning time in an air bag or blanket restraining device intended to protect people in automobile accidents is less than one second while warning time for certain industrial hazards is based upon annual x-ray examination.

The rows of Figure 5-9, i_1, i_2, . . ., are derived from time-line analysis of the system. Each value of i represents a discrete interval of time in the life cycle in which a potential hazard has been identified. Here, too, the number of intervals and the amount of time each represents depend upon the nature of the system and its life cycle phase. In some instances the time of concern may be very brief, such as the maximum Q portion of the ascent phase of a spacecraft; in other cases, the time of concern may be lengthy, such as the deployment interval of an atomic submarine.

The values in the matrix, the E_{ij}'s, represent the probability that action taken by personnel, automatic sensing and controlling equipment, or some combination of both will avoid injury or damage during the ith interval of the system life cycle, provided the amount of warning time specified in the jth row is available. In some instances it may be necessary to replicate Figure 5-9 three times, establishing three sets of acceptable levels for E_{ij}, each set corresponding to one of the three hazard classifications, I through III. It may be desirable, at least in the case of Class I hazards, to conduct a sensitivity analysis to determine how the hazard classification varies from I to III as a function of warning time and what costs are incurred in the system by increasing the warning time. In some instances, it may be possible to transform a Class I hazard into a Class II hazard by adding fractions of a second to the warning time. In other instances, at least for Class I hazards, it may be desirable to establish estimates of the E_{ij}'s during earliest iterations of the hazard analyses and to characterize these as inadequate, marginal, or adequate. Sensitivity analyses and characterizations of hazards are obtained more readily through fault tree analysis than through hazard mode and effects analysis. The procedure for performing fault tree analysis is described in Chapter 7.

5.13 SCOPE OF THE HAZARD ANALYSIS

The scope of the hazard analysis conducted for any given system is controlled by time and funding limitations. Such limitations should be explicitly stated or derivable from the contract establishing the safety effort to be carried out in support of the program. Typical qualitative requirements established for complex systems often include constraints such as the following:

1. No single failure shall be cause for a fatality.
2. All potential failures which can cause extensive damage to the system shall be made fail-safe by design.
3. All hazardous substances, components, and operations in the system shall be isolated from any other substances, components, and operations with which they are incompatible.
4. The system shall be designed so that injury to personnel and damage to equipment are minimized in the event of an accident.
5. Undue exposure of personnel to physiological or phychological stresses which might cause errors leading to mishaps shall be avoided by design and procedure.

Implementation of these and other qualitative criteria through specific constraints is based upon the character of each system. A set of quantitative constraints that may be used to control the scope of the hazard analysis for a system follows:

1. Class III-inducing hazards are permitted to have a probability of occurrence equal to or less than P_i.
2. Class II-inducing hazards are permitted to have a probability of occurrence equal to or less than P_j, where $P_j < P_i$.
3. Class I-inducing hazards are permitted to have a probability of occurrence equal to or less than P_k, where $P_k < P_j$.

These constraints, in conjunction with a set of qualitative constraints established for the system, control the scope of the hazard analysis because of their effect upon the design and use of the equipment required for hazard avoidance. If, for example, one of the constraints established explicitly or by contract implication requires the avoidance of fatalities brought about by single failures, then it is necessary to provide hardware or functional redundancy for each system element whose failure could cause fatality. In addition, it is necessary in this instance to provide sensing and switching mechanisms so that downtime resulting between the time a system element required for life support fails and its hardware or functional redundancy is brought on-line is sufficiently small.

Although by definition all P_k's that relate to Class I hazards are less than P_j's that relate to Class II hazards, there can be wide variation within each group. For example, in the case of a nuclear weapon system the value of P_k for inadvertent launch or detonation will be much smaller than the values of P_k for other types of Class I hazards associated with this system. The combination of qualitative and quantitative constraints also determines the accuracy required of the hazard analysis and, consequently, the number of iterations needed for its completion. Since the engineering and analytical portions of a hazard analysis become more complex and costly with each successive iteration, the analysis needs to be structured so as to (1) determine early in the program those system design features which need to be modified in order to eliminate or control hazards and (2) avoid complex analysis of events whose occurrence is unlikely or would not, if they occurred, cause serious hazards.

5.14 SURVIVAL RATE

Fundamentally, the preliminary hazard analysis is qualitative, intended as an aid for preventing and controlling hazards. However, as a preliminary analysis is transformed into an intermediate hazard analysis by successive iterations, it tends to become quantitative. When this occurs, it may become feasible to add one more factor and obtain a preliminary estimate for the ability of a system's users and operators to survive a hazard's occurrence.

Survivability, defined in Section 3.2.5 as the degree to which a system will withstand the environment in which it is placed and not suffer abortive impairment, can be measured in several ways. One such measure, of obvious concern, relates to the number of fatalities, f, to be expected per unit of system use. Typical units of use include a trip by any means of transportation, a given interval of work for an individual or an entire facility, and a stay in a hospital for specified surgery. Of equal concern is W, the number of times per unit of system use that survival rather than fatality occurs subsequent to the occurrence of a hazard. It is possible to define W either in terms of survival under the circumstance that no injury is incurred, or in terms of survival even though the hazard results in an injury. The first measure, f, implies that occurrence of a hazard is synonymous with fatality. The second presumes that there is sufficient warning time, a reliable abort system, and, if necessary, reliable rescue operations. For a specific hazard, H, W may be written as

$$W = EHR \tag{5.1}$$

where

$E =$ the probability that escape is possible

$H =$ the number of times that the specific hazard is expected to occur per unit of system use

$R =$ the reliability of the abort and/or rescue system

Since the hazard being considered is expressed as a probability, the value of H is equal to the number of times H occurs divided by the total number of trials. For example, if $P(H)=0.001$, H can be expected to occur, on the average, once per thousand units of system use. If, for a specific circumstance, $E=0.90$, $H=0.0005$, and $R=0.95$, substituting these values into Equation 5.1 yields $W=4.3=10^{-4}$ survivals per hazard occurrence per unit of system use. Since it is assumed that occurrence of a hazard with no escape potential is synonymous with fatality, the worst case number of fatalities is equal to the value of H, 5×10^{-4}, per unit of system use. The expected number of fatalities per unit of system use is, therefore,

$$H-W=(5\times10^{-4})-(4.3\times10^{-4})=0.7\times10^{-4}$$

If it is more convenient to consider abort and rescue separately, Equation 5.1 may be written

$$W=EHR_aR_r \tag{5.2}$$

where R_a and R_r are the reliabilities of the abort and rescue systems, respectively. It should be noted that R and, consequently, W may be a function of time, because R_a or R_r may be time-dependent. This dependency, if it exists, can take a variety of forms. Under certain conditions it is reasonable to expect that the abort capability of a system remains constant in time. The ability of an automobile to pull over to the side of the road as a consequence of engine failure does not, on the average, change significantly during its lifetime. For other systems the abort capability could decrease linearly or exponentially as a function of time. The ability of an airplane to abort after the takeoff roll has been initiated decreases inversely and nonlinearly as ground speed increases.

In either instance, the ability to successfully carry out abort is usually enhanced as the amount of warning time is increased. Since E, the probability of escape, is, among other things, a function of warning time, and R is a function of the system time line, it is possible to prepare a matrix analogous to Table 5.7 which presents W_{ij} or f_{ij} as matrix elements. In this instance, each matrix element is the expected value, in relation to a specific hazard, for the warning time and for that portion of the system time line indicated by the chart. Here, too, it may be desirable to determine the sensitivity of the matrix elements to changes in warning time and to use these data as a design constraint. The total number of fatalities per unit of system use may be obtained for any given warning time by summing the f_{ij}'s in that row for the set of matrices required to describe all potentially fatal hazards.

5.15 VARIATIONS IN THE HMEA

There are a variety of special purpose hazard modes and effects analysis (HMEA). Four of particular note are

1. Preliminary hazard analysis (PHA)
2. Fault hazard analysis (FHA)
3. Operating hazard analysis (OHA)
4. Subsystem hazard analysis (SHA)

The format offered earlier (Figure 5-7) may be varied somewhat to orient each analysis more precisely to its specific purpose, but their appearances are similar. Their differences, discussed next, are in the purpose each serves.

5.15.1 Preliminary Hazard Analysis

The preliminary hazard analysis (PHA) is concerned with tradeoff studies that are important in the early phases of system development or, in the case of an operational system, facilitate an early determination of state of safety. The PHA output is usually used for

1. Developing system safety requirements
2. Preparing performance and design specifications
3. Establishing the framework for other hazard analyses that may be performed

5.15.2 Fault Hazard Analysis

The fault hazard analysis (FHA), is performed later in the system development cycle. The FHA addresses each identifiable component in the system with two questions:

1. How can this component fail?
2. What are the system or subsystem consequences of this failure?

5.15.3 Operating Hazard Analysis

The operating hazard analysis (OHA) focuses on hazards that result from tasks, activities, or operating system functions that occur as a system is stored, transported, or exercised. OHA provides a basis for

1. Design changes to eliminate hazards to provide safety devices and safeguards
2. Procedures for servicing, training, handling, storage, and transportation of the system
3. Publication of warning, caution, special instructions, or emergency procedures for operations

4. Identification of the timing or operations or system functions that relate to hazardous occurrences

5.15.4 Subsystem Hazard Analysis

The subsystem hazard analysis (SSHA) identifies

1. Subsystem design features or components that can create hazards
2. Interface considerations between system elements that can create hazards
3. Subsystem information that facilitates integration of the full system hazard analysis

Additional Reading

Directorate of Aerospace Safety, USAF. *System Safety Hazard Analysis.* Norton Air Force Base, Calif.: U.S. Air Force, 1966.
———. Report, USAF-Industry System Safety Conference, Appendices M and O, 25–28 February 1969.
Invisible Industry. *Wall Street Journal,* 14 July 1969, p. 1.
Redding, R. J. Intrinsic Safety, London: McGraw-Hill, 1971.
Robb, D. A., Philo, H. M. and Goodman, R. M. *Lawyer's Desk Reference.* Rochester, N.Y.: Lawyer's Co-Operative, 1968.

Chapter 6

FAULT TREE ANALYSIS: DRAWING THE TREE

6.1 HISTORY OF FAULT TREE ANALYSIS

The development of fault trees as they are known today in the domain of system safety is generally ascribed to A. B. Mearns of Bell Laboratories. When developed, the procedure was intended as a logical technique for describing flow in data processing equipment, an obvious concern to Bell Labs. The application of logic tree technique to system safety studies occurred when it was observed that methods used to describe the flow of logic in data processing equipment were also applicable to describing the flow of undesirable events in a system. It was further observed that a logic tree lends itself readily to quantifying the probability that any discernible event could occur.

Formal introduction of logic trees into the public domain as fault trees occurred about 1961 and is generally ascribed to the Boeing Company. Boeing's original application was to provide a safety evaluation of the Minuteman Launch Control System, to prevent occurrence of inadvertent detonation or launch of a nuclear device. The consequences of either occurrence are so severe that unlikely events, such as a squirrel gnawing through an intersite cable and causing a missile launch, were considered in the analysis.

At first, fault tree analysis for system safety studies was used mainly for atomic energy programs, space travel, and complex military programs that were not severely restricted monetarily. Since then, fault tree usage has expanded by orders of magnitude. In part, this expansion is due to the need now felt by so many industries and technologies for intensive safety analysis, a need brought about to some extent by government regulations and a litigious society. Fulfillment of this need has been helped by the introduction and growth of computer-aided fault tree analysis (CAFTA). First efforts in computer-aided fault tree technology were made by R. E. Vesely about 1964 at the Idaho Nuclear Corporation. At that time computers powerful enough to examine large fault trees were rare and even when they were available, the time required for exact solution was too long to permit anything but Monte Carlo procedures. Since then, improvements in computer speed and capacity, along with new languages and programming, have made computer-aided fault tree analysis generally available.

6.2 COMPARISON WITH HAZARD MODE AND EFFECTS ANALYSIS

The ideal safety analysis is done by combining the hazard mode and effects analysis (HMEA) and fault tree methodology so as to utilize the advantages inherent in both. The advantages and disadvantages of these two methodologies are discussed next; a procedure for combining them is discussed in Section 8.8.

6.2.1 The Hazard Mode and Effects Analysis

A hazard mode and effects analysis (HMEA), an inductive procedure, is conducted by hypothesizing failures in lower levels of a system and examining the effects such failures have on higher assemblies and, finally, on the system. For example, the consequences that occur when a resistor fails shorted, fails open, or drifts out of tolerance, too high or low, would be germane in an HMEA carried out at the detail-part level. An HMEA is thorough since it postulates all conceivable failures to every component in the system. In some instances the analysis reveals that a postulated failure in a component or subsystem does not cause a hazardous condition in the system. Such data may nevertheless yield information needed for reliability or maintainability studies. The HMEA has some inherent disadvantages:

1. Since detailed knowledge of system hardware is required, a complete HMEA can only be conducted during latter portions of the development cycle. This could make it difficult for safety to influence the design of the system during its formative stages. Some help may be derived from an HMEA conducted at the subsystem level, but the inherent value of an HMEA makes it more useful at lower levels.
2. Only single thread failures are examined by an HMEA. The safety analyst's concern is, however, with probability of hazard occurrence regardless of whether the cause is a single or multiple thread event.
3. The HMEA, because it is hardware-oriented, does not readily admit consideration of problems that are not caused by hardware difficulties.
4. The HMEA does not lend itself to precise quantification of probability of occurrence. It does, however, permit assessment of the criticality of the potential hazard in general terms such as possible loss of life, only damage to equipment, and negligible effects.

6.2.2 Fault Tree Methodology

A fault tree analysis, a deductive procedure, is begun by postulating an undesirable event, such as system failure or personnel injury. The tree is then developed downward from this *top* event without any limitations on the nature of the problems that could cause the top event to occur. A fault tree can include considerations such as the following.

1. Defects in design, material, or workmanship.
2. Errors of omission or commission in operations and maintenance tasks.
3. Occurrence of credible accidents.
4. Interfaces between the system under analysis, other systems, and the environment.
5. Occurrence of natural phenomena such as earthquakes and hurricanes.

Because it is developed downward from the top, a fault tree can be initiated during early stages of a program and updated as more information is developed. In addition, a fault tree allows for qualitative and quantitive analyses of single and multiple thread failures.

6.3 NATURE OF A FAULT TREE

Although fault trees per se are relatively new, the use of trees as a descriptive device is not. In its general form, a tree is a special representation of a *graph*. (The first known paper on graph theory, written by the Swiss mathemetician Leonhard Euler [1707–1783], was published in 1736 and was motivated by the now famous Königsberg bridge problem.) A graph is defined as

> A collection of points, called nodes or vertices, with connections between certain pairs of these nodes, called branches or arcs, links, or edges.

Figure 6-1 illustrates a graph in which the nodes are represented by circles, and the branches connecting them are represented by straight lines. A graph is considered to be a *network* if appropriate quantification is associated with each arc. For example, if Figure 6-1 is considered a model of airline routes between six cities, the mileage between each city pair is inappropriate quantification. Rather, in a fault tree, the quantification of concern is associated with flow, as, for example, the number of aircraft flying between each city pair.

Development of a definition for a fault tree requires introduction of the terms chain, path, and cycle. A *chain* between nodes i and j is a sequence of branches that connects nodes i and j. For example, the set of branches (1, 2) (2, 3), (3, 6), or the reverse of this sequence, describes one of the several chains shown in Figure 6-1 that connect nodes 1 and 6. If the chain is

Figure 6-1
Example of Connected Graph

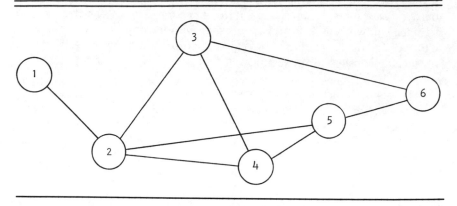

polarized so that the direction of travel along the chain is specified, it is called a *path*. A *cycle* is a special type of chain and is defined as a chain connecting node *i* to itself. In Figure 6-1 the chain (2, 3), (3, 4), (4, 2) is a cycle. A graph is considered to be connected if there is at least one chain connecting every pair of nodes. Therefore, Figure 6-1 is a connected graph. However, if branches (2, 4), (3, 4), (4, 5), and (5, 6) are removed, the figure is no longer a connected graph. There is, for example, no path between nodes 1 and 4 when these branches are missing. One way to determine whether a graph with no cycles is a tree is to count the number of nodes and branches. A theorem in graph theory states that a graph with n nodes is connected if it has $(n-1)$ branches and no cycles.

Table 6-1 presents some examples of connected graphs. In the first, airports form the nodes, airlanes make up the branches, and aircraft movement describes the flow; the second deals with transportation on the ground. A new connected graph could be created by combining the first two examples,

Table 6-1
Examples of Network-Type Connected Graphs

Nodes	Branches	Flow
Airports	Airlanes	Aircraft
Intersections	Roads	Vehicles
Switching points	Wires, channels	Messages
PERT events	Planned activities	Tasks
Valves	Veins	Blood

and trains, barges, and other forms of transportation could be added to form still another. The other examples in Table 6-1 illustrate that connected graphs are used to describe an extensive variety of activities. It will be shown that some of the examples in Table 6-1 are, or could be, trees, a restricted form of connected graphs.

6.4 FAULT TREE DEFINED

A *fault tree* may be defined as

> An oriented, connected, graph with no cycles, in which no more than one arc is directed out of the same vertex.

The term *arborescence* is used in graph theory for graphs so defined. That is, a connected graph becomes a tree when two additional restrictions are imposed: direction and openness. Direction requires that the orientation of flow associated with each branch be specified for both its input and its output. The second restriction, openness, is necessary to ensure that there are no closed loops. The need for these restrictions will become evident in the following discussion.

6.4.1 Orientation in a Tree

A definition of orientation involves the notion of capacity, the capability of a branch to support flow in a specified direction. *Capacity* is

> The upper limit of the feasible magnitude of the rate of flow or the quantity of flow, whichever is applicable, to each branch of the network.

Networks describing the flow of electrical energy or hydraulic fluid, for example, can have flow in either direction along a given branch. Such networks, however, involve the use of chains and, consequently, cannot be trees. A branch of a graph is considered to be oriented if the capacity for flow is equal to zero in one direction. There is, therefore, a sense of direction attributable to such a branch, so that a node at one end may be considered to be a point of origin and a node at the other end a destination. In a fault tree this quality of orientation establishes one group of nodes, the end events, as points of origin with flow emanating from them and converging toward the primary or top event. The capacity of any branch is related to the probability of flow occurring in that branch, which in system safety translates into the probability that a hazardous event will occur.

Consider the implications of the orientation restriction when applied to the connected graphs in Table 6-1. Roads can be structured to allow traffic to flow in either direction so that the network is a connected graph, or they can be restricted to one direction, thereby changing the network to a tree.

The valves controlling the flow of blood through veins model a tree under normal operating conditions. Flow in a backward direction, mathematically allowable if the model is a network, indicates a defect in an animal system.

6.4.2 Closed Cycle Considerations

Consider the implications of the restriction forcing closed cycles in a tree when applied to the connected graphs in Table 6-1. It is obvious that closed loops exist in some of the connected graphs. In fact, return to the starting point is expected in most transportation systems. Closure is, by definition, not permitted in a tree; the need for this restriction is discussed later.

6.5 PROCEDURE FOR PREPARING A FAULT TREE

A fault tree consists of a set of undesirable end events, the terminal nodes—whose occurrence is combined in accordance with a set of rules described by the nonterminal nodes (the logic gates) and whose probability of occurrence, calculated in accordance with a set of logical rules, determines the probability of occurrence of the top, undesirable event.

In general, a hazard analysis using fault tree methodology can be conducted by carrying out the following steps:

1. Obtain source materials needed for the analysis
2. Conduct a preliminary analysis
3. Construct a fault tree
4. Simplify the fault tree
5. Estimate probabilities of failure for end events
6. Identify areas in the system requiring corrective action

These steps are not necessarily sequential, and their general interrelationships are shown in Figure 6-2. Preliminary analysis of the system elements and the manner in which they are used yields a preliminary selection of the elements and procedures that may be contributors to the occurrence of the undesired event specified by the top fault tree event. Descriptions of these elements and procedures, together with their specifications, enable estimates of failure rates and hazard occurrences to be obtained.

This preliminary analysis provides the information needed to construct a first model of the fault tree, beginning with the top undesired event and showing all significant combinations of intermediate events that could cause system failure. The failure and hazard rates for the elements involved are then combined with the pertinent time-line data, obtained by engineering judgment or by examination of operating and maintenance instructions, to determine the probability of contributory failure. These probabilities are

Figure 6-2
Procedure for Fault Tree Analysis

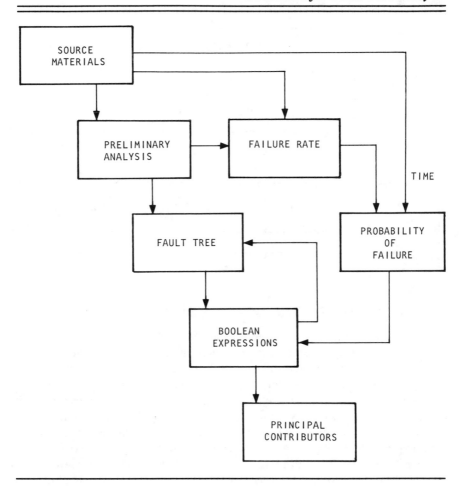

then used to highlight the principal contributors to the occurrence of the top-level undesired events.

The general form of the fault tree is depicted in Figure 6-3. In this, the flow of analysis is downward, and the development of the tree is continued until all the input fault events are defined in terms of basic, identifiable faults which may then be quantified for evaluation. This generalized description of the nature of a fault tree and the manner in which it is constructed are amplified in the next few sections.

Figure 6-3
General Form of a Fault Tree

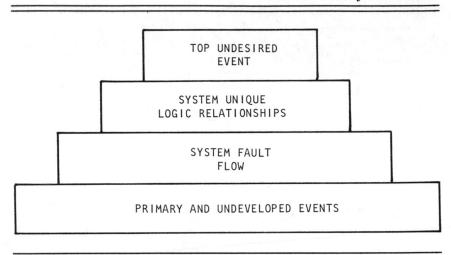

6.6 BASIC FAULT TREE COMPONENTS

Members of most professions enjoy the ability to communicate with one another by means of symbolic representations and terminologies that are relatively unknown outside their circle. Programmers use flow charts and computer languages, behavioral scientists make use of sociograms, and organic chemists utilize the graphic language of molecular change. One such communication used in system safety involves manipulation of nodes by means of fault tree methodology. The nodes in a tree may be categorized as terminal or intermediate. A given node in a network may be referred to as a *source* if every one of its branches is oriented so that flow moves away from that node and as a *sink* if the flow in each of its branches is oriented toward that node. In a fault tree, the terminal nodes act as sources, and the top event, the root, acts as a sink.

6.6.1 Terminal Nodes (End Events)

The term *end event* in fault tree terminology refers to a node in the fault tree, an input sample point to the tree, that may be transformed from a safe state to an unsafe state, thereby creating a source of flow. Consequently, the set of terminal nodes in a fault tree make up the $\bar{S}$ subset of

$$P(S) + P(\bar{S}) = 1$$

Therefore, this set of end events expressed the general case noted in Figure 6-3 including the four categories A–D that represent the possible combinations of equipment malfunction and human error (refer to Table 5-2). Each member of the set of end events is presumed to exist in a fault tree as long as there exists a reasonable probability that it can occur. A member ceases to exist if conditions for its existence are no longer satisfied, that is, if some form of corrective action or control is taken so that the probability of its occurrence is reduced to a negligible level. The set of end events used most often in constructing a fault tree is presented in Figure 6-4.

6.6.1.1 The Circle End Event

The circle describes a basic end event, a fault in the system, that requires no further development and acts as a source in a tree. It is understood when a circle is used that the frequency and mode of failure of the end event is available or can be estimated. For example, the fault event "relay contact sticks open" is a basic event whose frequency of occurrence may be estimated from data sources such as Military Handbook 217 (MIL-HDBK 217)— *Reliability Prediction of Electronic Equipment*—the Government Industry Data Exchange Program (GIDEP), manufacturer's data, or testing.

On occasions, variations of a solid-line circle are used to provide additional detail about the nature of a circle end event. For example, a broken circle is used sometimes to describe a basic fault event caused by a human act of omission or commission, and a double, striped circle is used to indicate failure to detect and correct a fault.

6.6.1.2 The Diamond End Event

The diamond, also referred to as an incomplete or undeveloped event, describes an event in a fault tree that is assumed to be basic even though further development is known to be possible. In some instances diamonds are used during early iterations of the tree before development of the system is complete and may subsequently be changed to circles when adequate information is available. In other instances, the development of the tree is terminated by the use of diamonds because information provided in the tree by adding further detail below the level of the diamond would not be useful

Figure 6-4
End Events

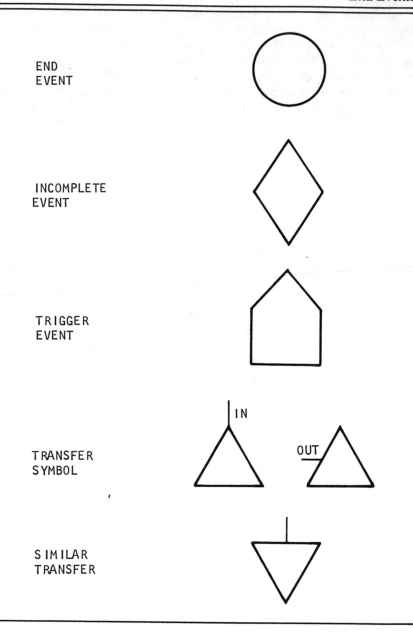

END
EVENT

INCOMPLETE
EVENT

TRIGGER
EVENT

TRANSFER
SYMBOL

IN

OUT

SIMILAR
TRANSFER

in the analysis. On occasion, a diamond is used to distinguish an end event whose flow is initiated by sources outside the system being modeled. In this latter use, it may lose its quality of incompleteness.

As in the case of the circle, variations of a solid-line diamond are sometimes used to present additional data about the nature of this end event. For example, the broken diamond is used in the same way as the solid one except that the fault events modeled are those which are due to personnel error. The double, striped diamond may be used to indicate failure of a human operator to detect and correct a fault. As in the cases of the plain and broken diamonds, this event is not developed any further through choice or because data are not available at that point in time.

A double diamond is sometimes used in a fault tree to describe results obtained from known causes that are not presented in the fault tree. This usually implies that the gates and end events not shown below the diamond do not make a contribution toward identifying hazards in the system.

6.6.1.3 The House End Event

The house event, also referred to as a trigger or switch event, indicates a situation that must occur in the system under normal operating conditions and is not considered to be a correctable fault. Examples include a phase change in a dynamic system, such as takeoff, flight, or landing of an aircraft, and changes in a human system caused by hormone secretions triggered by sudden fright or anger. When quantifying a fault tree, a trigger event is assigned a probability of occurrence equal to zero or one.

6.6.1.4 The Transfer Symbol End Event

The transfer symbol end event, represented by a triangle, is used as a shorthand notation to avoid the need for redrawing part of a tree. It is, therefore, not an end event that corresponds to a sample point whose occurrence initiates flow in the tree. Rather, the triangle acts as a pointer that permits the analyst to locate a specific section of the tree, one that needs to be duplicated and the duplicate moved elsewhere. The location of the section pointed out by the triangle can be elsewhere in the same tree or more remotely located as, for example, in another document. As shown in Figure 6-4, two types of transfer symbols are used in fault trees:

1. The transfer-in symbol, a triangle with a line emanating from its apex. This is used like an end event in that
    a. It appears at the bottom of a branch in a tree.
    b. It acts as a source for nodes at higher levels.
    When used, the transfer-in symbol is connected to a gate at the next higher level and points to another section of the tree that is marked by a transfer-out symbol. To show a tree in its entirety without the use of shorthand notation, the transfer-in symbol is replaced by the section referred to by the transfer-out symbol, and the transfer-out symbol is deleted.
2. The transfer-out symbol, a triangle with a line emanating from the center of one side, points to a section of the tree to be duplicated. In use, the line from the side of the triangle is connected to a gate. Everything below that gate is duplicated and used to replace the transfer-in symbol paired with the transfer-out symbol.

Transfer symbols may be nested. That is, a section pointed to by a transfer-out symbol can contain other transfer-in and -out symbols at lower levels. Concentrated use of this shorthand in the tree can make the overall tree more difficult to comprehend. It might be noted that this and some other shorthand descriptions discussed later are most useful when a tree is drawn by hand. With computer-aided fault tree technology a tree can be drawn in its entirety with little extra effort without transfer symbols, making the tree easier to examine and lending itself more readily to quantification.

6.6.1.5 Similar Transfer

The last end event symbol shown in Figure 6-4, similar transfer, is rarely used. The similar transfer symbol behaves like a conventional transfer symbol, but refers to a similar rather than an identical segment of the tree. Consequently, this symbol is useful when the differences between the two sections are known or can be readily determined.

6.6.2 Intermediate Nodes (Logic Gates)

The processes that control the way in which end events modeled by a fault tree are combined, the intermediate nodes of a fault tree, are called logic gates. Each logic gate in the tree, with one exception, has at least two arcs leading into it and exactly one leading out. An intermediate node acts like a sink in relation to arcs leading in and as a source for the one arc leading out. The capacity of the arc leading out of an intermediate node may be determined by some logical combination of the capacities of the arcs leading into that node. The logic gates used most often in constructing a fault tree are shown in Figure 6-5. The discussion about each, particularly the gates with qualifiers, also describes typical uses. Of the gates shown in Figure 6-5, the two top ones, AND and OR, are used almost exclusively in tree models; according to some purists, they are the only gates that should be used. This claim is based on three factors:

1. Gates requiring qualifiers, the bottom four in Figure 6-5, can also be expressed in terms of AND or OR gates. This may require a bit of ingenuity, but the result can represent a precise description of the system.
2. There are universally accepted expressions for describing flow through AND and OR gates that allow for standardized quantification of a tree. On the other hand, while gates with qualifiers can also be quantified, this can occur in various ways depending upon the user's interpretation of the system. Thus, quantification of gates with modifiers cannot be readily standardized.
3. The dual of a fault tree, obtainable by changing AND gates to OR gates and OR gates to AND gates, has significance. (See Section 7.6.) Since gates with qualifiers have no simply definable duals, the dual of a tree containing gates with qualifiers is not readily obtainable.

6.6.2.1 The AND Gate

The AND gate model shows flow coming in from the bottom through events $E_1, E_2, \ldots, E_n$. The output E can occur if and only if all the input events occur. In Boolean notation,

$$E = E_1 \cap E_2 \cap \ldots \cap E_n. \qquad (6.1)$$

Conversely, E will not occur if one or more of the $E_1, E_2, \ldots, E_n$ events does not occur. Table 6-2 presents combinations for two input events, a and b. In the table, H (high) corresponds to occurrence and L (low) corresponds to nonoccurrence of the input events a and b and the output event E.

Figure 6-6 illustrates the use of the AND gate to model an electrical circuit, Figure 6-6a. In this example, it is presumed that only the switch can be faulty and that the undesirable event E occurs when the light goes out. Since either a_1 or a_2 can close the circuit, both must be open for E to occur. The AND gate, therefore, describes the logic of this circuit and is modeled in Figure 6-6b. The rectangle contains information concerning the event resulting from the combined occurrence of fault events a_1 and a_2 and plays no part

Figure 6-5
Commonly Used Logic Gates

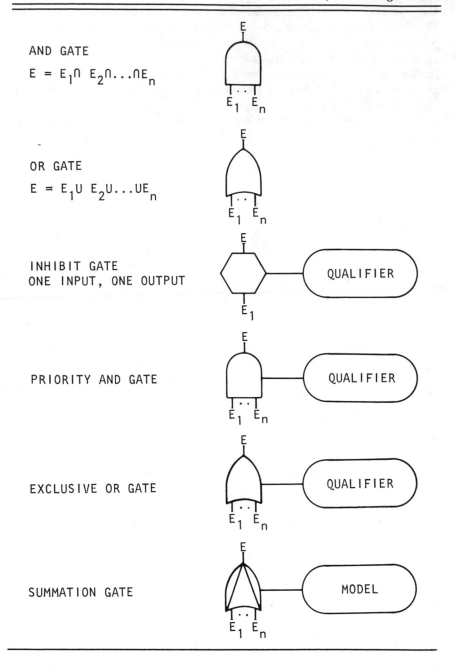

AND GATE

$E = E_1 \cap E_2 \cap \ldots \cap E_n$

OR GATE

$E = E_1 \cup E_2 \cup \ldots \cup E_n$

INHIBIT GATE
ONE INPUT, ONE OUTPUT

QUALIFIER

PRIORITY AND GATE

QUALIFIER

EXCLUSIVE OR GATE

QUALIFIER

SUMMATION GATE

MODEL

Table 6-2
Combinations of
Two Input Events
For an AND Gate

Inputs		Outputs
a	*b*	*E*
L	L	L
L	H	L
H	L	L
H	H	H

Figure 6-6
AND Gate Fault Tree for Switches in Parallel

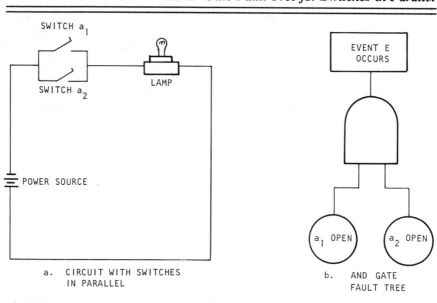

a. CIRCUIT WITH SWITCHES
IN PARALLEL

b. AND GATE
FAULT TREE

in describing the logic by which events are combined in the tree. The circle is used in the figure to indicate that no further attempt is made to develop the fault identified as "open switch."

6.6.2.2 The OR Gate

In the case of an OR gate, the output E occurs if any one or more of the inputs $E_1, E_2, \ldots, E_n$ occurs. In Boolean notation,

$$E = E_1 \cup E_2 \cup \ldots \cup E_n. \tag{6.2}$$

Conversely, E will not occur if and only if all the $E_1, E_2, \ldots, E_n$ inputs do not occur. Table 6-3 presents combinations for two input events a and b. H and L have the same meaning given for the AND gate table of combinations.

Figure 6-7, a circuit with two switches in series and a corresponding fault tree, illustrates the use of an OR gate. As in Figure 6-6, the undesirable event occurs when the lamp goes out. Since switches a_1 and a_2 are in series, opening either one or both will cause the undesirable output event to occur.

Figures 6-6b and 6-7b describe mechanisms by which input events may combine so that an undesirable output event can occur. They do not, however, provide an estimate of the likelihood that the output event will occur. This information may be derived by obtaining probabilities for the occurrence of the various input events described by Equations 6.1 and 6.2. The probability of E occurring in an AND gate representation is given by

$$P(E) = P(E_1) \cap P(E_2) \cap \ldots \cap P(E_n) \tag{6.3}$$

and in an OR gate representation the probability of E occurring is expressed

$$P(E) = P(E_1) \cup P(E_2) \cup \ldots \cup P(E_n) \tag{6.4}$$

Procedures for determining probabilities for the occurrences of these input events and combining them to obtain a likelihood of occurrence for the output event in the general case are discussed in detail in subsequent chapters.

Table 6-3
Combination of
Two Input Events
for an OR Gate

Inputs		Outputs
a	b	E
L	L	L
L	H	H
H	L	H
H	H	H

6.6.2.3 The Inhibit Gate

The inhibit gate is one of the three shown in Figure 6-5 whose logic includes the use of a qualifier (or modifier). An oval is used in this text to represent a qualifier; regular and irregular hexagons may also be found in the literature to express this function. The purpose of the modifier is to provide greater precision to the fault tree representation of a system. As suggested earlier, this type of gate could add complexity to the quantification process.

Inhibit gate is often written with quotation marks around the word *gate* because its restriction to exactly one input makes this the exception to the rule that two or more inputs must enter each gate. The output *E* occurs in an inhibit gate model if both the input and the qualifier occur, but there is no specific way to describe in advance the method by which the two are related or combined. Examples of combinations in an inhibit gate include the use of a qualifier

1. Whose occurrence is normal to one or more system states or is the result of a hazardous occurrence.
2. Which, whether normally present in the system or occurring as a result of a hazard, exists for the entire life of the system or only during specific periods of time.

Figure 6-7
OR Gate Fault Tree for Switches in Series

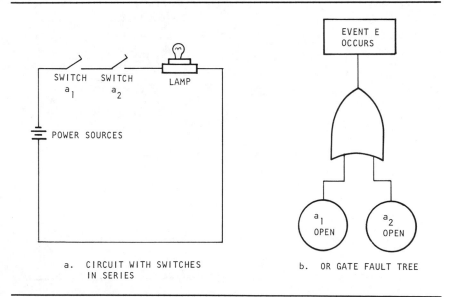

a. CIRCUIT WITH SWITCHES IN SERIES

b. OR GATE FAULT TREE

3. Which is related to the input of the inhibit gate in accordance with some known relationship or in a random fashion.

INHIBIT GATE ILLUSTRATED. Figures 6-8a and 6-8b illustrate two typical uses of inhibit gates. It should be noted that neither figure represents a complete section of a fault tree. To complete that section of the tree, the gates and events below the gates denoting "tire rupture" needs to be developed, and additional levels, arcs, and nodes need to be added between the inhibit modifier and the output event "accident occurs." In the figure the symbol ⌂ is used to denote a gate of any type.

The output event in Figure 6-8a can occur if and only if the input event occurs and the road is wet. One distinction between the application of this modifier and the use of a conventional AND gate is that the binary operations discussed thus far for describing an output event by Boolean notation do not suffice for specifying the output of the modifier in terms of the input event and the qualifying statement. Rather, it is necessary to use the notion of "conditional probability," which describes the probability that a certain event will occur given that some specific, related event has occurred. The subject of conditional probability is expanded in Section 8.4.

Figure 6-8
Inhibit Gates Illustrated

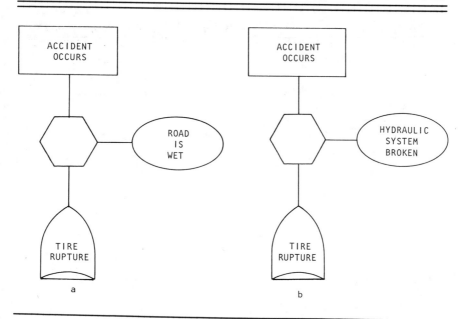

The output event shown in Figure 6-8b can occur if and only if the input event occurs and a break occurs in the hydraulic system. Both figures are similar in that the output event of each is presumed to be a certainty when both the input event and the qualifying condition occur. In addition, the probability of occurrence of the input event, tire rupture, is the same for both figures. However, the probability that the qualifying event will occur is described differently in each instance. In Figure 6-8a the probability that the qualifying event will occur can be expressed quantitatively as a function of known weather history for the geographical area under consideration. Consequently, one number will suffice for a given problem.

The probability that the qualifying event shown in Figure 6-8b will occur is related to such factors as the area of the hydraulic system, the strength of the materials used, and whether the lines in the hydraulic system are redundant. A quantitative assessment of this qualifier needs to be treated statistically as a stochastic (random) process.

THE STOCHASTIC PROCESS.   In a stochastic process, the determination of whether the output event "accident occurs" comes to pass requires an understanding of random phenomena and random events. The notions of stochastic processes, random phenomena, and random events introduced here are dealt with at greater length in Section 9.2. Briefly, a random phenomenon is one member of a set of occurrences whose outcomes under similar circumstances are not always identical but whose outcomes under similar circumstances do have statistical regularity. Therefore, a random event has a relative frequency of occurrence whose limiting value is the true probability of occurrence of the random event. Combinatorial logic in a fault tree that requires the use of stochastic processes for determining probabilities makes use of a random condition qualifier.

It should be noted that occurrence of the input event "tire rupture" when the qualifier does not occur may or may not represent a hazard. This suggests that a given input fault event may be presented in more than one place in a fault tree. In the case of Figures 6-8a and 6-8b, it may be presumed that another branch of the tree contains the event "tire rupture" as an input event, but without qualification. "Tire rupture" is related at that location to one or more output events through a suitable set of arcs and nodes.

6.6.2.4 Priority AND Gate

The priority AND gate, shown in Figure 6-5, performs the same function as a conventional AND gate except that the inputs must occur in the manner stipulated by the modifier. As in the case of an inhibit gate, there are a variety of ways in which this can be specified. Some examples are given in Figures 6-9 through 6-14.

Figure 6-9
Sequential AND Gate

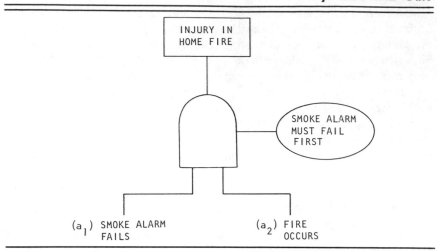

SEQUENTIAL PRIORITY AND GATE.   Figure 6-9 illustrates the case in which occurrence of the output event requires that (1) both input events a_1 and a_2 occur and (2) event a_1 occurs before event a_2. There is no limit to the number of inputs which may be sequentially ordered by a priority AND gate. One of the several feasible generalizations is shown in Figure 6-10. In this example, the output event is presumed to occur (1) if all the a_n input events occur and (2) event a_i occurs before a_j.

The case in which all input events needed to cause flow must occur simultaneously is usually a trivial application of the sequential modifier. If the operation of the system is such that only the normally trivial case need be considered, then an AND gate suffices as a model. If the trivial case needs to be considered along with the general cases, then the tree could be drawn as shown in Figure 6-11. In the example presented in Figure 6-9, the simultaneous occurrence of both a fire and failure of the smoke alarm system is not trivial but, with one exception, is extremely unlikely. The exception is a design such that a fire could start in the alarm system. Inherent safety with regard to this hazard is obtainable by using batteries instead of household 115V supply.

COMBINATORIAL PRIORITY AND GATE.   A priority AND gate may also be qualified as shown by the combinatorial modifier in Figure 6-12. In this figure the fault event occurs when any two of the a_n events occur. In the general case, a combinatorial priority AND gate may be used to express the

Figure 6-10
Sequential AND Gate—General Case

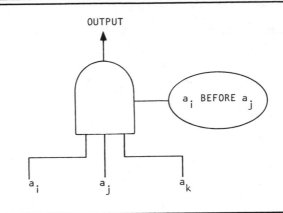

Figure 6-11
OR Gate with Two Priority AND Gates

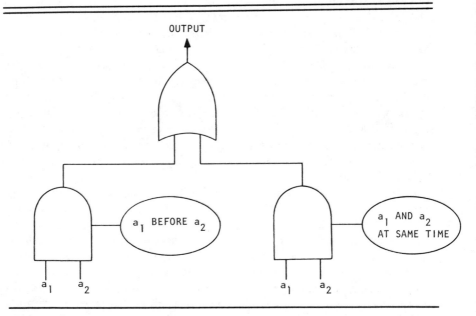

Figure 6-12
Combinatorial Priority AND Gate

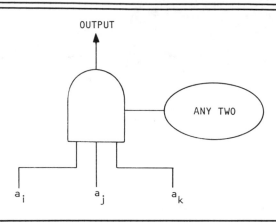

occurrence of an output from a priority AND gate with m inputs when $n = 2$, $3, \ldots, m - 1$ are required to occur.

The sequential and combinatorial modifiers can be combined. Substituting a combined modifier in place of the one used in Figure 6-12 establishes the occurrence of an output fault event if and only if any two of the a_i, a_j, a_k events occur, and if the event that occurs first is the one with the lower of the two subscripts of the a_i's that do occur. In the general m input case, the number of inputs that must occur may be set to $2, 3, \ldots, m - 1$, and any logical statement, such as less than, greater than, or no two adjacent subscripts, may be used in the qualifier.

It was noted earlier that gates with qualifiers can often be modeled solely by the use of AND and OR gates. Figure 6-13, for example, is equivalent to the combinatorial priority AND gate shown in Figure 6-12 when $n = 3$.

HAZARD DURATION MODIFIER. In addition to qualifying an AND gate by restricting the number of input events or the order in which they occur, a qualifier may be used to control the occurrence of the output event as a function of hazard duration time. The use of this qualifier implies that all the input events must occur in order for the output event to occur, as in the case of any AND gate. The fundamental use of a hazard duration modifier is to describe the relationship between input fault events and output hazard events when elimination or modification of the input fault event within a specified period of time prevents occurrence of the hazardous output. Consider the case in which an individual is injured so as to cause bleeding. The

Figure 6-13
Equivalent for Combinatorial Priority AND Gate

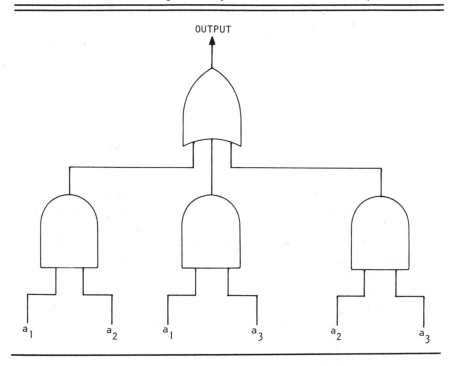

hazard duration modifier describes the length of time between occurrence of the input event and the occurrence of the output event "death." Repairs effected in less than the amount of time specified in the modifier will diminish the classification of the output event from Class I to II or III.

In the case where redundant equipment exists in a system to provide some function needed for survival, the hazard duration modifier may be associated with the length of time required to switch from one equipment to another or with the length of time required for repair. As an example of this relationship, consider the case where some installation, such as a hospital, has a standby source of electricity for use in the event of primary power failure. There is some interval of time between failure of the primary power source and start-up of the secondary source, less than the time specified by the hazard duration modifier, which could cancel or modify the effect of the input fault event. In some cases, the input fault event is considered to be cancelled if the repair or switching time is less than the hazard duration time. It might be noted, however, that in this circumstance the vulnerability of the system to power failure is increased until such time as the primary power supply is put back into operation. Effecting repairs does not imply that the input fault event, although repaired, cannot subsequently recur during the time span of concern.

The hazard duration qualifier may also be used with inhibit gates, discussed in Section 6.6.2.3, and with the exclusive OR gate, discussed in the next section.

6.6.2.5 Exclusive OR Gate

An exclusive OR gate has an output if and only if flow occurs in any one of the inputs and in none of the others. Combinations for an exclusive OR gate with inputs a and b are shown in Table 6-4. H (high) corresponds to occurrence and L (low) corresponds to nonoccurrence of the input events a or b or the output event E.

Figure 6-14 illustrates one use of an exclusive OR gate. The figure represents a portion of a transportation system which is presumed to have two thrusting devices mounted symmetrically about the system centerline, such as occurs in some hardware and biological systems. The undesirable event "asymmetrical thrust" can occur in response to the input events only if the qualifier "not both simultaneously" modifies the OR gate inputs. The simultaneous occurrence of both input events does not result in the output event shown in Figure 6-14. Rather, the output event in this circumstance is loss of thrust, which may or may not be a hazardous occurrence, depending upon the nature of the system.

6.6.2.6 Summation Gate

The summation gate, the last one shown in Figure 6-5, is used to model the case in which a set of input events can each cause the output event to occur or can combine with one or more of the other inputs to cause the output to occur. Figure 6-15 illustrates the use of a summation gate for the output event "smog alert" when there are exactly three inputs, a, b, and c, reactive hydrocarbons, nitrous oxide, and ozone, respectively.

The use of a summation gate implies several consequences. First there is a level for each of the three inputs which by itself can cause the output E to occur. Let these amounts be A_a, A_b, and A_c, for inputs a, b, and c, respectively. Second, there are an infinite number of values for a, b, and c, taken two at a time, that can cause E to occur. Consider a and b. There is an amount $A_a^1 < A_a$ that when combined with an amount $A_b^1 > 0$ that can cause E to occur. The amount A_a^1 can decrease continuously from a value slightly less than A_a to a value A_a^n that is slightly greater than zero. For any amount of reactive hydrocarbons A_a^k, $A_a > A_a^k > 0$, there is a corresponding amount of nitrous oxide A_b^j, $A_b > 0$ such that the combination of A_a^k and A_b^j causes E to occur. As the value of A_a^k approaches zero in this infinite set of pairs, the

Table 6-4
Combinations for
an Exclusive OR
Gate

Inputs		Output
a	b	E
L	L	L
L	H	H
H	L	H
H	H	L

value of A_b^j approaches A_b, the relationship between the two being expressed by the model in the qualifier. In like fashion, there is an infinite set of values for combinations of pairs *a* and *c* and *b* and *c* and for the trio *a*, *b*, and *c*.

6.7 FAULT TREES ILLUSTRATED

The manner in which the logic gates and end events presented in Figures 6-4 and 6-5 are combined to construct a fault tree is discussed next and the subject of computer-drawn fault trees is introduced. Figure 6-16 utilizes all

Figure 6-14
Exclusive OR Gate in Asymmetrical Thrust

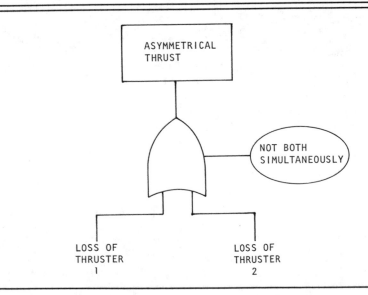

the gates and end events shown in Figure 6-4 and 6-5 to construct a tree. In addition to illustrating how gates and end events are connected to form a tree, this figure illustrates the use of computer technology in drawing fault trees. Some comments about the format used in computer-drawn fault trees will aid in interpreting Figure 6-16 and other trees presented in this text that are drawn in this fashion.

When drawn by computer, each gate, qualifier, and end event consists of two parts; the gate, qualifier, or end event as shown in Figures 6-4 and 6-5 and a rectangle above. The computer keeps track of the gates, qualifiers, and end events used in the tree by labels assigned to each node and qualifier. These labels need not appear in the tree; they are, however, shown in Figure 6-16. For example, the OR gate at the head of the tree is labeled TOP, and the end events in the tree are labeled E_1, E_2, . . . , E_8. The rectangle associated

Figure 6-15
Summation Gate Illustrated

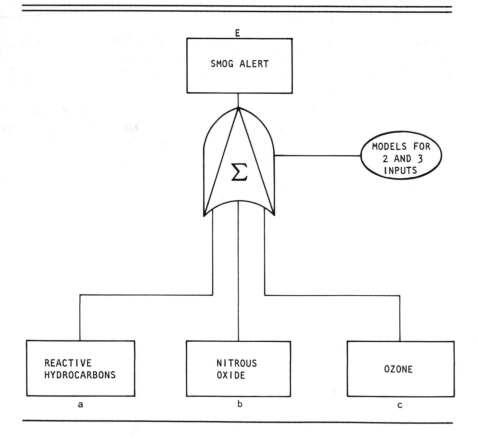

Figure 6-16
Computer-Drawn Fault Tree

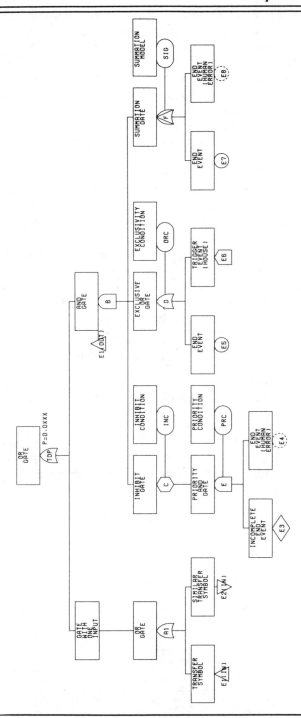

with each node and qualifier contains descriptive material associated with that particular node or qualifier. The rectangle does not describe the logic by which end events are combined nor does it aid in the quantification process. Consequently, the rectangle is optional when drawing a tree by hand. It may be included with all nodes or with some. When left out of a tree, the information placed in the rectangles drawn by computer may be placed inside the hand-drawn node or qualifier. Gate, qualifier, and end event labels are of no particular value in hand-drawn trees.

The fault tree shown in Figure 6-16 has the following characteristics:

1. Two OR gates, one AND gate, one inhibit gate, one exclusive OR gate, one summation gate, and one priority AND gate are used in the tree for describing the logic. Two complete end events, one incomplete end event, one trigger event, two transfer symbols (one in and one out), and one similar transfer symbol act as sources.
2. There is one point in the tree, located on the left-hand side below OR gate TOP and above OR gate *A*1, where a gate has exactly one input. While permissible, this is a trivial case in a tree since there is no distinction between AND and OR gates with exactly one input. (The computer is programmed not to draw a gate if it has exactly one input.)
3. OR gate TOP shows a "probability of occurrence" in this figure. There is no actual probability associated with this gate, but its presence is intended to indicate that preparation of a system model can be one step in a process that leads to quantification in terms of probability of occurrence. The process by which fault tree quantification is carried out is discussed later.

Figure 6-17
Sample Fault Tree

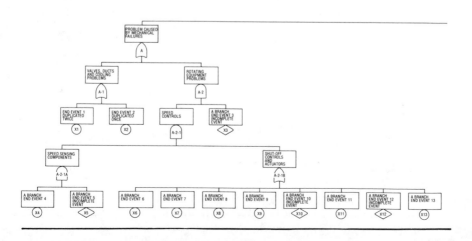

6.7.1 Typical Fault Tree

In appearance, Figure 6-17 is a typical fault tree. It represents no real system, and the text associated with each gate and end event is intended to be descriptive of a generalized tree rather than of a real system. The figure describes a possible hazard to the system, at OR gate TOP, through three subbranches

A.   Problems caused by mechanical failure
B.   Problems caused by electrical failure
C.   Maintenance problems

The output of exactly one AND or OR gate is written as shown in Equations 6.1 and 6.2. The output event in the more general case, one in which a fault tree utilized several AND and OR gates, is obtained by using the principles established in Section 3.3. That is, the binary operations union and intersection and the unary operation complement can be combined and manipulated through conventional algebraic techniques. The manipulations performable with these three operators are commutative, distributive, and associative. Further, the two identity elements, *I* and *W*, behave in relation to union and intersection in accordance with the expressions

$$A \cup I = I \tag{6.3}$$

$$A \cap W = A \tag{6.4}$$

Figure 6-17 (Continued)

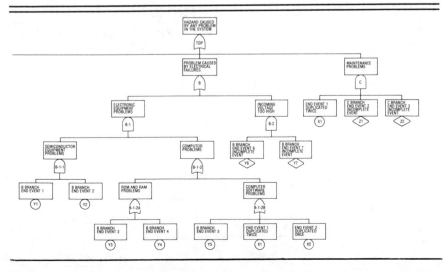

Applying these rules of combination to the nodes in Figure 6-16,

$$(\text{TOP}) = A \cup B \cup C \qquad (6.5)$$

where

$$A = (A\text{-}1) \cup (A\text{-}2) \qquad (6.6)$$

$$B = (B\text{-}1) \cup (B\text{-}2) \qquad (6.7)$$

$$C = (X1) \cap (Z1) \cap (Z2) \qquad (6.8)$$

Equation 6.8 cannot be further reduced since $X1$, $Z1$, and $Z2$ are end events, but A-1, A-2, B-1, and B-2 of Equations 6.6 and 6.7 can be further reduced. Expanding these to the next lower level they become

$$(A\text{-}1) = (X1) \cup (X2)$$

$$(A\text{-}2) = (A\text{-}2\text{-}1) \cap (X3)$$

$$(B\text{-}1) = (B\text{-}1\text{-}1) \cap (B\text{-}1\text{-}2)$$

$$(B\text{-}2) = (Y6) \cap (Y7)$$

Further expansion so that each logic node is expressed solely in terms of terminal nodes is left to the reader.

6.7.2 Residential Fire Alarm System

Consider a residential fire alarm system. The major components of such a system and the manner in which they are related are shown in Figure 6-18. A top-level, partial fault tree model of this system is shown in Figure 6-19,

Figure 6-18
Residential Fire Alarm System

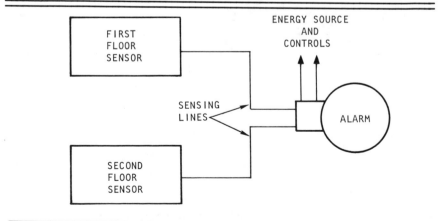

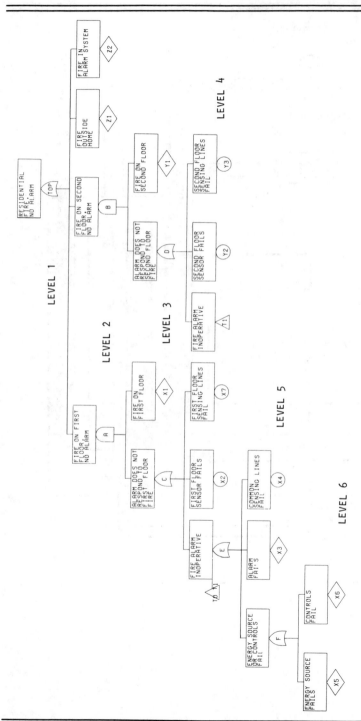

Figure 6-19
Residential Fire Alarm Fault Tree

Figure 6-20
Domestic Hot-Water Heater

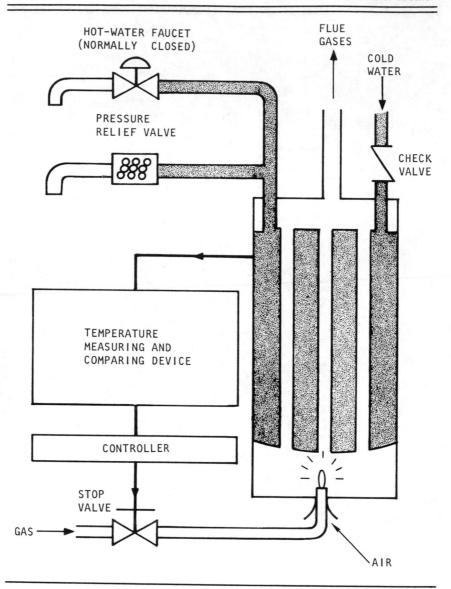

HOT-WATER FAUCET
(NORMALLY CLOSED)

FLUE
GASES

COLD
WATER

PRESSURE
RELIEF VALVE

CHECK
VALVE

TEMPERATURE
MEASURING AND
COMPARING DEVICE

CONTROLLER

STOP
VALVE

GAS

AIR

which uses six levels to trace the activities from the top, undesirable output event to the lowest fault events. Starting from the top, each successive lower level approaches the lowest level inputs by one logical step per level change.

The two major branches, *A* and *B*, reflect fires on the first and second floor of the residence, respectively. Since the alarms on both floors are identical, it is possible to terminate one selection with the transfer-in symbol *T*1. The development of the transferred material is shown below OR gate *E*. Both diamonds and circles are used as end events in the *A* and *B* branches. The circles indicate basic end events requiring no further development, while the diamonds imply that further development is required.

The other two branches relate to fire occurring outside the home or in the alarm system. Lest the reader consider this latter factor unimportant, note that one manufacturer of smoke detectors agreed to pay a $100,000 civil fine for alleged failure to notify the government that about 115,000 of its detectors could overheat and catch fire. Completeness in fault trees, a limiting factor in their usefulness, is discussed in Section 6.9.

6.7.3 Residential Hot Water Heater

A second illustration utilizes the typical domestic hot water heater, shown schematically in Figure 6-20. A top-level fault tree for the undesired event "tank rupture" is presented in Figure 6-21. In this example, it is noted that tank rupture can occur either as a result of defects in the tank or through external sources. The former is represented by branch LEV2-*B*, and three representative problems are shown by end events $Y1$, $Y2$, and $Y3$. The latter is developed under OR gate LEV2-*A*. In this external overpressure, problems caused by natural phenomena, such as earthquakes or hurricanes, are shown as a transfer from some other, unknown location. Typical causes of internal overpressure are shown below OR gate *A*.

6.8 ADDITIONAL LOGIC GATES

The AND and OR gates and gates with modifiers described thus far, although sufficient for drawing fault trees, are not the only ones used for this purpose. Other variations, developed to model specific, perhaps unique, processes by which events can combine to cause undesirable events or to act as a shorthand representation of AND or OR gates, can be found in the literature. Two examples are presented in this section.

6.8.1 NOT Gate

The NOT gate is illustrated in Figure 6-22 with two inputs; it simulates the case where the output fault event occurs if and only if a_1 but not a_2 (written $\bar{a}_2$) occurs, where a_1 and a_2, may be any type of input.

Figure 6-21
Hot-Water Heater Fault Tree

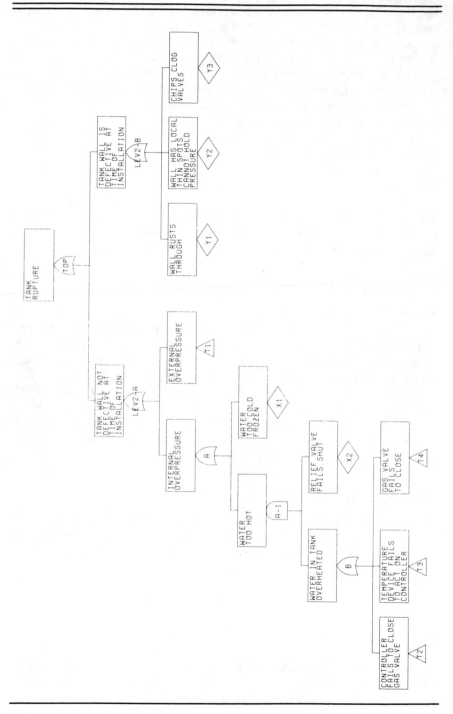

Figure 6-22
NOT Gate

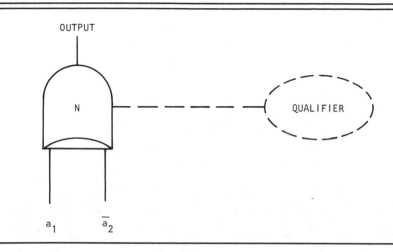

The symbol for the NOT gates shows that an AND or OR gate may be used, and the dashed qualifier indicates that its use is optional. The use of the NOT modifier can be illustrated by an example in public health in which the output event is "becomes ill with flu virus." The set of input events that contribute to the occurrence of this output event includes exposure to the virus and qualities of the human system which offer resistance to the illness. The NOT input event is "inoculation with given viral strain." The output can occur, therefore, if a_1 (exposure) occurs and a_2 (inoculation) does not occur. The hazards associated with receipt of the inoculation may be shown else-where in the tree. The qualifier is not needed for the case illustrated by Figure 6-22. It is, however, needed to express models such as

a_1 or a_2 or . . . or a_n, and not a_i
a_1 or a_2 and . . . and a_n, and not a_i
a_1 or a_2, and not a_3, and a_1 before a_2

The NOT gate has another use in conjunction with the exclusive OR gate. This is discussed in the next chapter.

6.8.2 Matrix Gate

In some cases it may be possible to simplify the form of a fault tree by using a special gate to represent a portion of the tree which otherwise would require several gates and associated modifiers. The matrix gate is one such abbrevi-

Figure 6-23
Matrix Gates

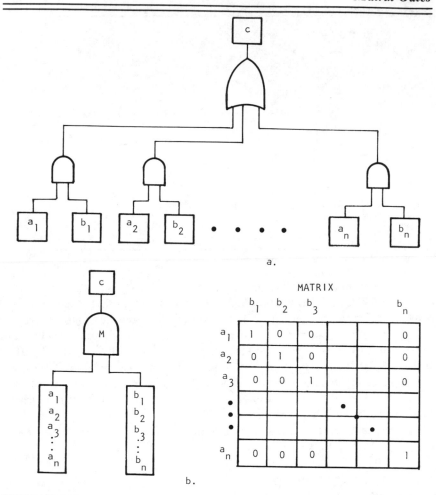

a.

b.

ated representation. The manner in which a variable matrix gate may be used in lieu of n AND gates, each with two inputs, and one OR gate is shown in Figure 6-23. The output event c is given by the expression

$$c = (a_1 \cap b_1) \cup (a_2 \cap b_2) \cup \ldots \cup (a_n \cap b_n) \qquad (6.9)$$

for Figure 6-23a and for its matrix gate equivalent, Figure 6-23b.

An illustration of a matrix gate application is presented by Figure 6-24. The figure represents a four-wire cable in which

Figure 6-24
Equivalent Fault Tree

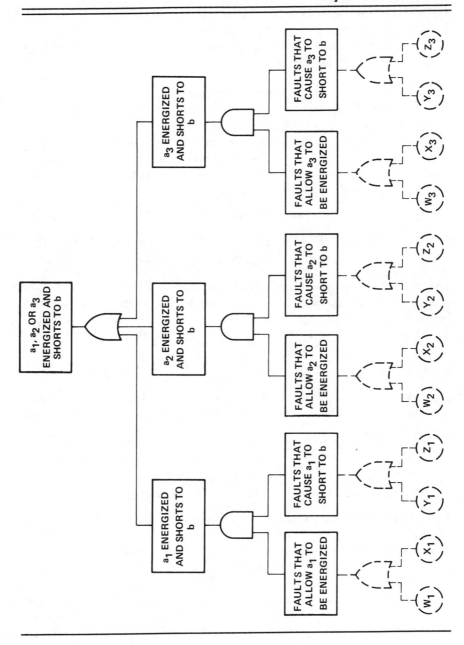

1. Wires a_1, a_2, and a_3 are energized at discrete time intervals.
2. Wire b is connected to an ordnance device.
3. The undesired output event occurs when any of the a_i wires is shorted as a result of an input fault event.

Although the a_i wires may differ in the function each performs, they may be modeled in the same fashion. One possible description of the basic input events that can occur in order to cause the output event to occur is shown in Figure 6-24 by the dashed portion of the fault tree. The equivalent tree in matrix gate form is shown in Figure 6-25. The dashed portion of the tree in Figure 6-25 is equivalent to the dashed portion of Figure 6-24. The expression for the output event, E_o, in both instances is

$$E_o = (a_1 \cap b) \cup (a_2 \cap b) \cup (a_3 \cap b) \qquad (6.10)$$

where the general term in Figures 6.24 and 6.25 $(a_i \cap b)$ is given by

$$(W_i \cup X) \cap (Y_i \cup Z_i) \qquad (6.11)$$

The savings in preparation time obtained by using a matrix gate rather than conventional logic gates are not significant for the example presented in Figure 6-24. If, however, a 50-wire cable is being represented by a similar matrix configuration, the savings in preparation time become significant.

6.8.3 Sampling Gate

Figure 6-26 represents a sampling gate. In this example all combinations of the n inputs are taken m at a time. The restriction that exists on the configuration is

$$1 < m < n \qquad (6.12)$$

6.9 FAULT TREE GENERATION, A SUMMARY

The information discussed thus far about the kinds of gates and end events used in constructing a fault tree plus the several illustrations is sufficient to allow the reader to begin constructing fault trees. Acquiring expertise in this skill, like that in a foreign or computer language, requires practice as well as an understanding of the methodology involved. One caveat is suggested: since the goal of fault tree construction is to model system conditions that can result in the top, undesired event, a thorough understanding of the system is a prerequisite for constructing a tree. It is recommended, therefore, that a system description always be made a part of the documentation describing a fault tree.

Figure 6-25
Matrix Gate Illustrated

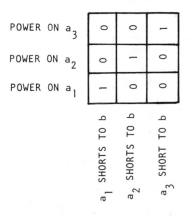

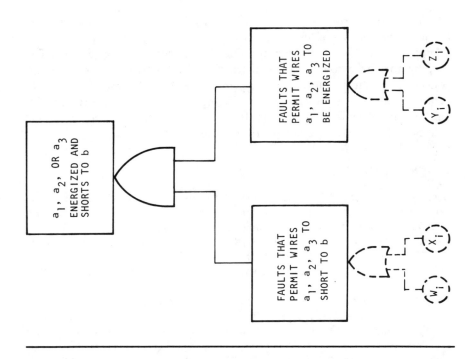

Figure 6-26
Sampling Gate

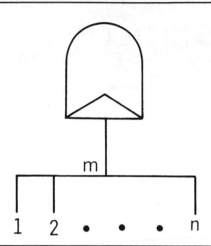

The basic shortcoming of fault tree analysis, of particular concern when the potential for loss of life needs to be considered, is uncertainty about the completeness of the tree. Computer-aided fault tree technology can help free investigators from routine tree construction and analysis, thereby allowing concentration on the more subtle aspects of system behavior. Because such subtleties are always present when modeling a system, a computer can offer aid but cannot replace the analyst. It can provide assurance, however, that routine failure modes are not overlooked in the study.

Additional Reading

Berge, C. *The Theory of Graphs and Its Applications.* New York: Wiley, 1962.

Browning, R. L. Loss control: Potential vs. probability. *Chemical Engineering,* October 20, November 17, December 15, 1969; January 26, 1970.

Directorate of Aerospace Safety, USAF. USAF-Industry System Safety Conference, 25–28 February 1969. Norton Air Force Base, Calif.: U.S. Air Force.

Dresher, M. Games of Strategy: Theory and Applications. Englewood Cliffs, N.J.: Prentice-Hall, 1961.

Ford, L. R., Jr. and D. R. Fulterson. *Flows in Networks.* Princeton, N.J.: Princeton University Press, 1962.

Frank, H., and I. T. Frisch. Network analysis. *Scientific American,* July 1970, pp. 94–103.

Haasl, D. F. Advanced Concepts in Fault Tree Analysis. Seattle: System Safety Symposium, June 8–10, 1965.

Mearns, A. B. Fault Tree Analysis: The Study of Unlikely Events in Complex Missile Systems. Seattle: System Safety Symposium, June 8–10, 1965.

Military Standard Graphic Symbols for Logic Diagrams. MIL-STD-008065. Prepared by the U.S. Navy, February 1962.

Minieka, E. *Optimization Algorithms for Networks and Graphs.* New York: Dekker, 1978.

Vajda, S. An Introduction to Linear Programming and the Theory of Games. London: Methune; New York: Wiley, 1960.

Chapter 7

QUALITATIVE FAULT TREE ANALYSIS

7.1 FAULT TREE AS A MODEL

A description of a fault tree, its nodes, the connections between them, and the nature and capacity of flow in its arcs is in actuality a model of a system. This one-to-one correspondence between fault tree and system allows predictions to be made of system behavior—specifically, the effect that changes in a tree's end events can have on the system. Perhaps the most important purpose of the model is to determine when flow in the terminal nodes can cause flow in the top event. This can be done by determining the *minimal cut sets* that make up the tree, a qualitative process. Quantitative considerations are discussed in Chapter 9.

7.2 MINIMAL CUT SETS DEFINED

A cut set is a group of end events whose collective occurrence in a tree ensures occurrence of the top event. Consequently, the collective failure of the analogous elements in a system is necessary and sufficient for system failure. A minimal cut set is a smallest set of end events such that if all of them are in a failed condition, the system is in a failed condition. Because it is a smallest set, one minimal cut set cannot contain another minimal cut set. Therefore, the set of minimal cut sets of a system is the (finite) collection of all unique minimal cut sets of that system. Consequently, a fault tree can be represented in nonredundant fashion by the union of all the minimal cut sets of the system.

A minimal cut set of size one in a tree corresponds to a single-point failure of the system; a minimum cut set of size two corresponds to a dual-point failure in the system; and so on. A minimal cut set is said to be in a failed state if, and only if, all its components are simultaneously failed. Consequently, a system is in a failed state if, and only if, one or more of its minimal cut sets is in a failed state. The probability that a system will fail is equal to the probability that one or more of the system minimal cut sets will fail.

The minimal cut sets of a system are, therefore, of prime importance since they describe the modes by which a system fails. In addition, they lend themselves to quantifying the likelihood that such failure will occur, for the minimal cut sets of a system depend only upon the structure of the fault tree and are invariant with respect to changes in component failure and repair distribution.

7.3 MINIMAL CUT SETS ILLUSTRATED

Examine Figure 6-17 from the bottom upward for minimal cut sets. Suppose end event $X4$, located on the bottom on the left-hand side, fails. This causes flow to occur to AND gate $(A-2-1A)$ but no farther since the output of gate $(A-2-1A)$ is given by the expression

$$(A-2-1A)=(X4)\cap(X5).$$

Therefore neither $X4$ nor $X5$ is single-point minimal cut set. Suppose, however, that $(X4, X5)$ is presumed to be a dual-point minimal cut set. This implies the occurrence of event $(A-2-1A)$. Following this flow along the arc to the next higher level gate shows that this presumption is false since the expression for flow out of AND gate $(A-2-1)$ is

$$(A-2-1)=(A-2-1A)\cap(A-2-1B).$$

It can be seen, therefore, that a smallest possible minimal cut set must contain three or more elements. It turns out in this example that four elements are required. It is left for the reader to verify that end events $X3$, $X4$, $X5$, $X6$ constitute a minimal cut set, that is, that their simultaneous failure causes event TOP to occur in the model and the system to fail. Strictly speaking, the minimal cut set should be written $(X3\cap X4\cap X5\cap X6)$. A formal procedure for finding minimal cut sets of any size in a tree is described in section 7.4.2.

Consider the end event $X1$, two levels above $X4$ and slightly to the right in Figure 6-17. Since flow out of OR gate $(A-1)$ is given by the expression

$$(A-1)=(X1)\cup(X2),$$

$X1$ could be a single-point minimal cut set. Moving higher in the tree, the expression for flow out of OR gate A is given by

$$A=(A-1)\cup(A-2).$$

Therefore, the occurrence of $X1$ ensures occurrence of A, and, since

$$(\text{TOP})=A\cup B\cup C,$$

occurrence of $X1$ ensures occurrence of TOP. End event $X1$ is, therefore, a single-point minimal cut set.

7.4 CUT SET ALGORITHMS

It is not generally possible to determine by inspection all the minimal cut sets in a tree containing hundreds of gates and events, but it is relatively simple to determine the minimal cut sets of size one. This ability to determine single-point cut sets is partially the reason analyses are oriented toward finding only single-point failures. However, while it is generally true that cut sets of size one have a greater probability of occurrence than cut sets of size two, and cut sets of size two have a greater probability of occurrence than cut sets of size three, and so on, exceptions could be costly. An algorithm—a rule of procedure for determining minimal cut sets in a tree regardless of size—is therefore preferable to minimal cut set determination by inspection. An algorithm allows all minimal cut sets to be determined automatically when a computer is employed. Pruning, that is, eliminating from further consideration those cut sets not considered significant to the outcome of the top event, may then be done in terms of probability of minimal cut set failure without regard to minimal cut set size.

A variety of algorithms have been devised for finding minimal cut sets, the general evolution being oriented toward decreasing the amount of computer time required while increasing the tree size that can be analyzed. Almost all these algorithms can be classified as bottom-to-top or top-to-bottom.

7.4.1 Bottom-to-Top Algorithm

A bottom-to-top examination of a fault tree for minimal cut sets is conducted somewhat as in Section 7.3 for informal minimal cut set determination. The algorithm is relatively straightforward. Flow is assumed to occur in one end event at a time, and the tree is examined to determine whether the top event occurs as a result of such flow. The set of end events whose failures, taken one at a time, cause failure of the top event is composed of the set of single-point minimal cut sets of the tree and is removed from the end events to be considered next. In the next iteration, permutations of the remaining end events are taken two at a time. Each pair is then checked to determine whether its failure causes the top event to occur. The set of pairs of end events that cause system failure is made up of the two-element minimal cut sets of the tree and is removed from further consideration. In the next iteration, all permutations of the remaining end events are taken three at a time, and the examination is repeated to find those that are minimal cut sets of size three.

The procedure can be continued until all minimal cut sets are found, or it can be terminated at some preselected minimal cut set size. Searching for minimal cut sets of sizes greater than are actually needed will indicate whether all minimal cut sets in the tree have been found. For example, if only cut sets of sizes one, two, and three are sought, it may nevertheless be useful to know whether the tree also contains cut sets of size four or higher.

This algorithm is relatively easy to program, but it may require prohibitive amounts of computer time, as the following typical scenario illustrates. Suppose there are 100 end events in a tree of which 5 are single-point failures, 10 are members of two-element cut sets, and the remainder are members of minimal cut sets of size three or greater. Suppose further that 0.01 sec is required for each test of the tree made by the computer for system failure. Then:

1. The selection of the 5 single-point failures would take $100 \times 0.01 = 1.0$ sec.
2. The selection of the two-element cut sets would take $95 \times 94 \times 0.01 = 89.3$ sec.
3. Continuing on to determine the three-element cut sets would require $85 \times 84 \times 83 \times 0.01 = 5926.2$ sec.

The total time for determining one-, two-, and three- element cut sets would then be about 1.7 hr, generally a prohibitive amount of computer time. The same task of a larger tree would not only require more passes by the computer, but would require more time per pass.

7.4.2 Top-to-Bottom Algorithm

The algorithm for determining the minimal cut sets of a fault tree by a top-to-bottom procedure is based on the fact that an AND gate in a tree increases the size of a cut set while an OR gate increases the number of cut sets. To describe this algorithm, the notion of Boolean Indicated Sets (BIS) is introduced. The BIS are defined so that if all the primary (end) events in a tree are different, the BIS are equal to the minimal cut sets. Like the bottom-to-top algorithm, this does not limit the fault tree to primary events that appear only once in the terminal nodes. The BIS are obtained by reduction from a set of Boolean Indicated Cut Sets (BICS). To obtain the BICS, each gate in the tree is assigned a value, w, and each primary event is assigned a value, f. The following definitions are applicable to the equations used to determine the BICS:

$$V_{w,i} = i^{th} \text{ input to gate } w$$
$$N_w = \text{number of inputs to gate } w$$
$$x = \text{the } x^{th} \text{ BICS}$$
$$y = \text{the } y^{th} \text{ entry in a BICS}$$

$D_{x,y}$ = a variable representing the yth entry in the xth BICS
$x$max = largest value of x used thus far
$y$max = largest value of y used thus far

The values w, f, $V_{w,i}$, N_w and the gate type, AND or OR, are input, and values of $V_{w,i}$ can be determined from observing values of w or f. $D_{1,1}$ is the first set equal to the value of w established for the gate connected to the TOP event. The process involves elimination of all values of w in the $D_{x,y}$ matrix. This is accomplished by locating w values in the $D_{x,y}$ matrix, noting values of x, y, and w and setting

$$D_{x,y} = V_{w,1}. \tag{7.1}$$

When w is a value for an AND gate,

$$D_{x,y\max+1} = V_{w,n} \text{ for } n = 2, 3, \ldots, N_w \tag{7.2}$$

where $y$max is incremented when n is incremented. When w is a value for an OR gate,

$$D_{m\max+1,m} = D_{x,m} \text{ for } m = 1, 2, \ldots y\max$$
$$m \neq y$$
$$= V_{w,n} \qquad m = y \tag{7.3}$$

where $x$max is incremented when n is incremented. The process described by Equations 7.1 to 7.3 is repeated until all the entries in the $D_{x,y}$ matrix become values of f, thereby determining the BICS. In general the BICS so determined are not minimal, so reduction to BIS is needed.

7.4.2.1 BICS Determination Illustrated

The process of determining BICS can be performed on the B branch of the fault tree in Figure 6-17. The $D_{x,y}$ matrix has the form shown in Matrix 7-1.

Matrix 7-1

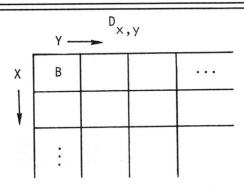

Equations 7.1, 7.2, and 7.3 are used to resolve the matrix as follows. First, *B* is resolved into Matrix 7-2.

Matrix 7-2

$$D_{x,y}$$

Y →

X ↓

B1		
B2		

(*B*-1) is then resolved into Matrix 7-3.

Matrix 7-3

y → $D_{x,y}$

x ↓

B-1-1	B-1-2	
B-2		

(*B*-1-1) is then resolved into Matrix 7-4.

Matrix 7-4

y → $D_{x,y}$

x ↓

Y1	Y2	B-1-2
B-2		

(*B*-2) is resolved into Matrix 7-5.

Matrix 7-5

Y1	Y2	B-1-2
Y6	Y7	

$D_{x,y}$ (y →, x ↓)

Columns 1 and 2 are now written in terms of end events (*B*-1-2) is resolved into its two inputs (*B*-1-2*A*) and (*B*-1-2*B*) (Matrix 7-6). Since their equivalent, (*B*-1-2), was associated with end events *Y*1 and *Y*2, each of the two new terms retains those end events. The number of lines in the *X* direction is, therefore, expanded to three. In the next step (*B*-1-2*A*) is resolved into its constituents, end events *Y*3 and *Y*4 (Matrix 7-7).

Matrix 7-6

Y1	Y2	B-1-2A
Y1	Y2	B-1-2B
Y6	Y7	

$D_{x,y}$ (y →, x ↓)

Matrix 7-7

Y1	Y2	Y3
Y1	Y2	Y4
Y1	Y2	B-1-2B
Y6	Y7	

$D_{x,y}$ (y →, x ↓)

The $Y1$ and $Y2$ terms associated with (B-1-2A) are repeated for $Y3$ and $Y4$, expanding the number of lines in the Y direction to four. In the final step (B-1-2B) is resolved into its component end events, $Y5$, $X1$, and $X2$ (see Matrix 7-8). The last line of the matrix, ($Y6$, $Y7$), remains unchanged since it is not reducible.

Matrix 7-8

$$D_{x,y}$$

$$y \longrightarrow D$$

Y1	Y2	Y3
Y1	Y2	Y4
Y1	Y2	Y5
Y1	Y2	X1
Y1	Y2	X2
Y6	Y7	

Each row of the final matrix (Matrix 7-8) is a Boolean Indicated Cut Set (BICS). These are examined to determine whether

1. Duplicate end events exist in any row
2. Any one BICS is a subset of another
3. Duplicate rows exist in the matrix

Any or all of these three will occur if end events are duplicated, an allowable condition in a fault tree. A BICS which includes another BICS is known as a superset, a rough inverse of a subset. A duplicate row in the matrix can be thought of as an improper superset, which is analogous to the notion of an improper subset. Elimination of duplicate end events and supersets is necessary to transform the BICS into minimal cut sets, also known as Boolean Indicated Sets (BIS). This process is illustrated in Section 7.4.2.2. Since none of the three conditions listed above exists in the final $D_{6,3}$ matrix, the B branch presented in Figure 6-17 is fully resolved. It contains one cut set of size two and five of size three.

1. *Y6 Y7*
2. *Y1 Y2 X1*
3. *Y1 Y2 X2*
4. *Y1 Y2 Y3*
5. *Y1 Y2 Y4*
6. *Y1 Y2 Y5*

The reader should check to see whether the union of all six minimal cut sets yields the entire *B* branch of Figure 6-17.

Reduction of the *A* branch proceeds as shown in Matrix 7-9.

Matrix 7-9

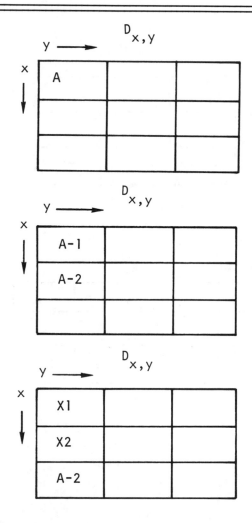

$$D_{x,y}$$

y $\longrightarrow$

x $\downarrow$

X1		
X2		
A-2-1	X3	

$$D_{x,y}$$

y $\longrightarrow$

x $\downarrow$

X1		
X2		
A-2-1A	A-2-1B	X3

$$D_{x,y}$$

y $\longrightarrow$

x $\downarrow$

X1			
X2			
X4	X5	A-2-1B	X3

$$D_{x,y}$$

y ⟶

x
↓

X1			
X2			
X4	X5	X6	X3
X4	X5	X7	X3
X4	X5	X8	X3
X4	X5	X9	X3
X4	X5	X15	X3
X4	X5	X11	X3
X4	X5	X12	X3
X5	X5	X13	X3

Consequently, the *A* branch has two cut sets of size one and eight cut sets of size four. There are no cut sets of sizes two, three, or greater than four.

7.4.2.2 BICS Reduction to BIS

The procedure by which the Boolean Indicated cut sets are reduced to Boolean Indicated Sets may be illustrated by assuming resolution of a tree has taken place, yielding Matrix 7-10.

$D_{4,3}$

y →		
1	2	
2	2	3
1	1	4
2	0	3
2	3	

x ↓

First, duplicates in each row are eliminated. This yields Matrix 7-11.

$D_{4,3}$

y →		
1	3	
2	4	
1	4	
2	4	3
2	3	

x ↓

Duplicate rows are eliminated next, yielding Matrix 7-12.

$$D_{4,3}$$

y →		
1	2	
2	3	
1	4	
2	4	3

x ↓

Supersets are eliminated next. In this example, the three-element BICS (2, 4, 3) contains no element that is not contained in another BICS and may therefore be eliminated. The resulting minimal cut sets are (1, 2), (2, 3), and (1, 4).

7.4.3 Minimal Cut Sets of Figure 6-17

The minimal cut sets leading to gates A and B of the sample fault tree in Figure 6-17 have already been determined. The C branch of this figure is small, and it can be seen by inspection that this branch contains one three-element minimal cut set, $(X1, Z1, Z2)$. The matrix showing all minimal cut sets of each branch of this figure is presented in Figure 7.1. The minimal cut sets of each branch have been numbered for reference. The 10 sets of the A branch, for example, are numbered $A1, A2, \ldots, A10$. The cut sets leading to the gate TOP, shown in Figure 7-2 can be found by using the top-to-bottom algorithm described in the previous section or by combining the cut sets found at the second level, that is, those sets leading to gates A, B, and C.

This notion of combining minimal cut sets at lower levels is introduced for two reasons. First, it is often true that minimal cut sets of small size are of interest, for example, when the branches at lower levels are relatively independent of one another. In such a case unsafe failure modes noted in one branch can be resolved without interfacing with those that may exist in other branches. In some instances the major branches of a fault tree model the efforts of different contractors, each concerned with their equipment and its failure modes. Once the minimal cut sets have been determined, it may be helpful to prune those whose occurrence is relatively unlikely. In Figure 7-1, for example, it might be reasonable to prune minimal cut sets of size four.

Figure 7-1
Matrix of Cut Sets Leading to A, B, and C Branches

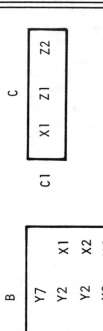

C

| C1 | X1 | Z1 | Z2 |

B

B1	Y6	Y7	
B2	Y1	Y2	X1
B3	Y1	Y2	X2
B4	Y1	Y2	Y3
B5	Y1	Y2	Y4
B6		Y2	Y5

A

A1	X1			
A2	X2			
A3	X4	X5	X6	X3
A4	X4	X5	X7	X3
A5	X4	X5	X8	X3
A6	X4	X5	X9	X3
A7	X4	X5	X10	X3
A8	X4	X5	X11	X3
A9	X4	X5	X12	X3
A10	X4	X5	X13	X3

Figure 7-2
Cut Sets for Gate TOP

```
******************************************************************
            MINIMAL  CUT   SETS  FOR  GATE  TOP
******************************************************************

     CUT  SETS  WITH    1  COMPONENTS

                1)    X1
                2)    X2


     CUT  SETS  WITH    2  COMPONENTS

                1)    Y6        Y7


     CUT  SETS  WITH    3  COMPONENTS

                1)    Y1        Y3        Y2
                2)    Y1        Y5        Y2
                3)    Y1        Y4        Y2


     CUT  SETS  WITH    4  COMPONENTS

                1)    X4        X3        X6        X5
                2)    X4        X3        X7        X5
                3)    X4        X3        X8        X5
                4)    X4        X3        X9        X5
                5)    X4        X3        X10       X5
                6)    X4        X3        X11       X5
                7)    X4        X3        X12       X5
                8)    X4        X3        X13       X5



     TOTAL  NUMBER  OF   CUT  SETS  FOUND  WAS       14

     ALL   CUT  SETS  HAVE  BEEN  DETERMINED,
```

This leaves the minimal cut sets of $A1$ and $A2$ in the A branch plus all others in the B and C branches. The process of combining cut sets that remain after pruning may be simpler than the process of finding all minimal cut sets to gate TOP.

The second reason for finding minimal cut sets to a top-level gate by combining those at lower levels is that large trees headed by an AND gate are sometimes not conveniently resolvable into component cut sets. The problem is that this task falls into the category of problems that are theoretically solvable but which require infinitely large computers or infinitely long computation time. This aspect is discussed further in Section 7.8 (see also Stockmeyer and Chandra in Additional Reading).

7.5 MINIMAL PATH SETS DEFINED

Path sets are, in the Boolean sense, duals of cut sets. Therefore, a path set is a collection of end events such that if all of them are functioning the system is functioning. A *minimal path set* is

> A smallest number of end events whose collective functioning is necessary and sufficient for the system to be functioning.

Because it is a smallest set, one minimal path set cannot contain another minimal path set. As in the case of minimal cut sets, the minimal path sets of a system is the (finite) collection of all unique minimal path sets for that system. A fault tree can therefore be represented in a nonredundant fashion by the union of all minimal path sets of the system.

A minimal path set is said to be in a nonfailed (operating) state when all of its components are simultaneously in a nonfailed state. Consequently, a system is in a nonfailed (operating) state if and only if one or more of its minimal path sets is in a nonfailed state. The probability that a system will remain in an operating state is therefore equal to the probability that one or more of its minimal path sets will remain in an operable state. The minimal path sets of a system describe the modes by which a system remains operable. In addition, they lend themselves to quantifying the likelihood of system operability for specific intervals of time, for the minimal path sets of a system depend only upon the structure of the tree and are invariant with respect to changes in component failure and repair distributions.

7.6 THE SUCCESS TREE

A tree made up of minimal path sets can be considered to be a success tree. It can be developed like a fault tree by selecting a suitable top event and constructing the success tree downward from that point or by the simple expedient of taking the dual of the fault tree. That is, changing all the AND gates to OR gates and changing all the OR gates to AND gates in a fault tree yields a success tree. In such a change the dual of each fault end event and gate is transformed into a success end event or gate. The dual of an end event "resistor fails" is "resistor does not fail," and the dual of a top event "accident occurs" is "accident does not occur."

The minimal cut sets of the dual of a fault tree are the minimal path sets of the original fault tree. Consequently, the algorithms described for cut set determination need not be modified to determine a tree's path sets. Computer-aided fault tree technology allows cut sets, path sets, or both to be determined in one run. If both minimal cut and path sets are known, it is possible to describe the modes by which

1. System failure can occur. Corrective action is taken as needed as described in earlier chapters.
2. System success can be assured. Operations and maintenance can put this information to good use.

The path sets for Figure 6-17 are presented in Figure 7-3. These sets conform to the general pattern that path set sizes are larger than cut set sizes. The reader is advised to select the *A* or *B* branch of Figure 6-17, create the dual,

Figure 7-3
Path Sets for Figure 6-17

```
******************************************************************************
                    MINIMAL PATH SETS FOR GATE TOP
******************************************************************************

    PATH SETS WITH   5 COMPONENTS
            1)    Y1      X1      X4      X2      Y6
            2)    Y2      X1      X4      X2      Y6
            3)    Y1      X1      X3      X2      Y6
            4)    Y1      X1      X5      X2      Y6
            5)    Y2      X1      X3      X2      Y6
            6)    Y2      X1      X5      X2      Y6
            7)    Y1      X1      X4      X2      Y7
            8)    Y2      X1      X4      X2      Y7
            9)    Y1      X1      X3      X2      Y7
           10)    Y1      X1      X5      X2      Y7
           11)    Y2      X1      X3      X2      Y7
           12)    Y2      X1      X5      X2      Y7

    PATH SETS WITH   7 COMPONENTS
            1)    Y3      X4      Y6      Y5      Y4      X1      X2
            2)    Y3      X3      Y6      Y5      Y4      X1      X2
            3)    Y3      X5      Y6      Y5      Y4      X1      X2
            4)    Y3      X4      Y7      Y5      Y4      X1      X2
            5)    Y3      X3      Y7      Y5      Y4      X1      X2
            6)    Y3      X5      Y7      Y5      Y4      X1      X2

    PATH SETS WITH  12 COMPONENTS
            1)    Y1      X1      X6      X2      Y6      X7      X8      X9     X10    X11
                  X12     X13

            2)    Y2      X1      X6      X2      Y6      X7      X8      X9     X10    X11
                  X12     X13

            3)    Y1      X1      X6      X2      Y7      X7      X8      X9     X10    X11
                  X12     X13

            4)    Y2      X1      X6      X2      Y7      X7      X8      X9     X10    X11
                  X12     X13

    PATH SETS WITH  14 COMPONENTS
            1)    Y3      X6      Y6      Y5      Y4      X7      X8      X9     X10    X11
                  X12     X13     X1      X2

            2)    Y3      X6      Y7      Y5      Y4      X7      X8      X9     X10    X11
                  X12     X13     X1      X2

    TOTAL NUMBER OF PATH SETS FOUND WAS    24

    ALL PATH SETS HAVE BEEN DETERMINED.
```

and find its cut sets. The path sets of the *A* and *B* branches of Figure 6-17 are presented in Figure 7-4. The comments made in Section 7.4.3 concerning the difficulties that may be encountered in determining minimal cut sets of large trees headed by an AND gate are valid for the case in which the dual fault tree is being analyzed; that is, when a search for minimal path sets in a large tree headed by an OR gate is undertaken.

7.7 CUT SETS FOR TREES WITH MODIFIED GATES

The algorithms—informal, visual perhaps, or formal—for finding cut or path sets of a fault tree presume that only AND and OR gates exist in the tree. This suggests a need for a method for transforming gates with qualifiers

Figure 7-4
Path Sets for Gates A and B

```
**********************************************************************
                  MINIMAL PATH SETS FOR GATE A
**********************************************************************

PATH SETS WITH   3 COMPONENTS

        1)    X1      X4      X2
        2)    X1      X3      X2
        3)    X1      X5      X2

PATH SETS WITH  10 COMPONENTS

        1)    X1      X6      X2      X7      X8      X9     X10     X11     X12    X13

TOTAL NUMBER OF PATH SETS FOUND WAS     4

ALL PATH SETS HAVE BEEN DETERMINED.

                  **********************************************************************
                  MINIMAL PATH SETS FOR GATE B
                  **********************************************************************

PATH SETS WITH   2 COMPONENTS

        1)    Y1      Y6
        2)    Y2      Y6
        3)    Y1      Y7
        4)    Y2      Y7

PATH SETS WITH   6 COMPONENTS

        1)    Y3      Y6      Y5      Y4      X1      X2
        2)    Y3      Y7      Y5      Y4      X1      X2

TOTAL NUMBER OF PATH SETS FOUND WAS     6

ALL PATH SETS HAVE BEEN DETERMINED.
```

into a configuration of AND and OR gates. Since the qualifiers are not always defined in a unique fashion, this cannot always be done automatically. A technique that serves this purpose in computer-aided fault tree technology presumes a set of transformations that is usually correct. Variations from these standards must be handled on an individual basis. A set of transformations considered suitable for automatic modification is shown in Figures 7-5 to 7-9.

7.7.1 Inhibit Gate

The transformation for the inhibit gate, shown in Figure 7-5, presumes that both the single input to the gate and the inhibit condition Q must occur for flow to be initiated out of the gate. Consequently, this transformation changes an inhibit gate to an AND gate with two inputs.

The simplicity of the transformation leads one to question the need for the inhibit gate. Would not its equivalent—an AND gate with two inputs— serve as well? It does serve as well for finding minimal cut and path sets. It does, however, add a dimension to the fault tree not otherwise possible. In Figure 6-8a, for example, the node leading into the inhibit gate is a fault associated with the system, while the qualifier describes an external factor.

7.7.2 Qualified AND Gate

The transformation for AND gates with qualifiers is shown in Figure 7-6. This results in a minimal cut set of size $N + 1$, where N equals the number of inputs to the original gate. This transformation is valid for sequential gates because all inputs and qualifiers must occur. The transformation does not work for the combinatorial priority AND gate in Figure 6-13.

Figure 7-5
Inhibit Gate Transformation

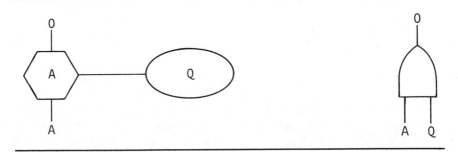

Figure 7-6
Qualified AND Gate Transformation

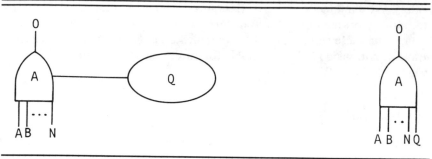

7.7.3 Exclusive OR Gate

The transformation for an exclusive OR gate is shown in Figure 7-7. For generality, Figure 7-7a has a conventional end event, an AND gate, and an OR gate leading to exclusive OR gate A. The transformation, Figure 7-7b, shows that NOT gates and NOT events are involved in the transformation. (The term $-i$ is used as the complement of i in the computer simulation in lieu of i to simplify programming.) NOT AND and NOT OR gates are defined in Figures 7-7c and 7-7d, respectively.

7.7.4 Summation Gate

The transformation for a summation gate is illustrated for three inputs in Figure 7-8 and is generalized later. The inputs and outputs of the summation gate shown on the left of this figure correspond to those discussed in Section 6.6.2.6 and shown in Figure 6-5. That is, the output corresponds to smog alert, and the inputs A, B, and C correspond to reactive hydrocarbons, nitrous oxide, and ozone, respectively. Figure 7-8 shows that the original summation gate is transformed into an OR gate with the same label, and several new gates are created:

1. A new OR gate labeled $N1$
2. Four new AND gates labeled $N2$, $N3$, $N4$, and $N5$

OR gate $N1$ indicates that events A, B, and C can each cause the output event to occur. There are, therefore, probabilities $P(A)$, $P(B)$, and $P(C)$ associated with inputs A, B, and C each of which can cause the output to occur. There are an infinite number of combinations of $A1$ and $B1$, where $A > A1 > 0$ and $B > B1 > 0$, such that flow occurs out of AND gate $N2$. For

Figure 7-7
Exclusive OR Gate Transformation

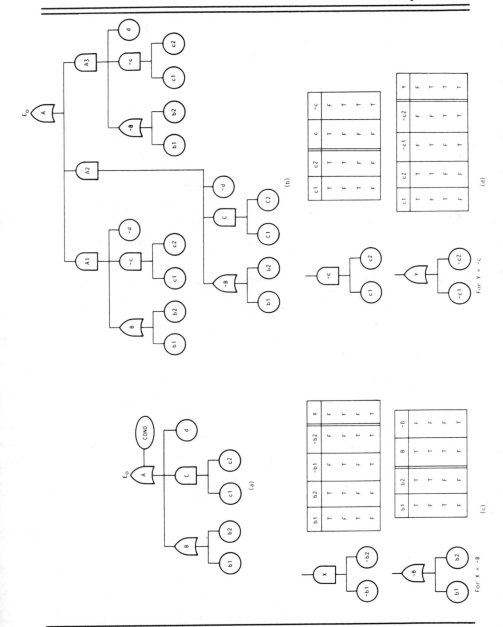

Figure 7-8
Summation Gate Transformation

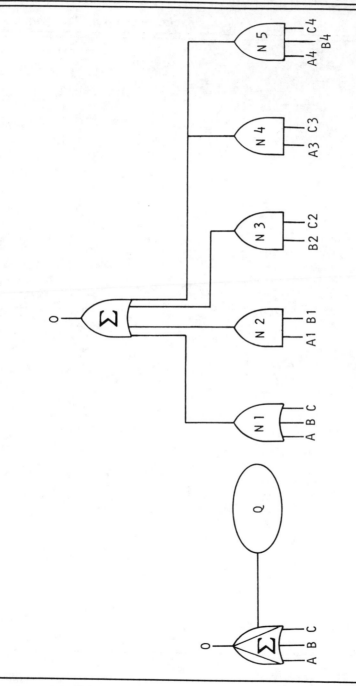

convenience one value is selected for each, so that values of $P(A1)$ and $P(B1)$ are established. An analogous situation exists for the other inputs to gates $N3$, $N4$, and $N5$.

The transformation in the proposed schema requires that each end event be unique. In Figure 7-8 this is done for event A by renaming it $A1$, $A3$, and $A4$ in each of its subsequent appearances. To understand why this is necessary, consider that each input has two facets. In this example, A represents a specific amount of reactive hydrocarbons, and there exists some probability that such an amount will occur. $A1$ refers to a lesser amount of reactive hydrocarbon which, in combination with $B1$, can give rise to the output. The amounts of reactive hydrocarbon and nitrous oxide that may combine to cause the output can vary, but one probability can be used to estimate that they will combine to cause the output. If $A1$ were not uniquely identified, then the cut set $(A1, B1)$ would not appear in the results, for the cut set $(A, B1)$ is a superset of the minimal cut set A which flows into OR gate $N1$.

It should be noted that the use of summation gates implies a large fault tree. If, for example, there are six inputs to a summation gate, the transformation yields 1 OR gate and 57 AND gates. The number of AND gates is derived from the expression

$$^6C_2 + {}^6C_3 + \ldots + {}^6C_6 = 57$$

where ^nC_r refers to the number of combinations of n dissimilar items taken r at a time. This is calculated by the expression

$$^nC_r = \frac{n!}{r!\,(n-r)!}$$

where

$$0! = 1$$

If it is assumed that there are exactly 10 gates and events leading to each input to the summation gate, then

1. The original summation gate has a total of 60 inputs.
2. Transformation according to the proposed schema results in the same 60 inputs plus 180 new end events.
3. If the inputs leading to event A in the original summation gate are replicated so that they also enter $A1$, $A2$, . . . , the new tree would contain the original 60 gates and end events plus 1800 new ones.

If, for example, there are five or six summation gates in a tree and the more precise representation is used, the resultant tree could have thousands of gates and end events—a situation that does not lend itself readily to precise analysis.

7.7.5 Sampling Gate

The transformation for sampling gates is illustrated in Figure 7-9 for three inputs and a sampling rate of two. The transformation is straightforward, but it should be noted that the number of new gates needed for the transformation increases rapidly as the number of inputs increases. Specifically, a sampling gate with m inputs taken n at a time requires p new gates where p is given by

$$p = \frac{m!}{n! \, (n-m!)}$$

7.8 TIME REQUIRED FOR CUT SET DETERMINATION

Minimal cut set determination, as noted in Section 7.4.3, can be a computational problem that requires a computer larger than any available and years of operation for solution. The time and computer core size required are functions of

1. The size of the tree, that is, the number of gates and end events in the tree
2. The syntax, the configuration of AND and OR gates in the tree

Of particular concern is the case in which a tree has four or more branches and is headed by an AND gate when minimal cut sets are of interest, or when it is headed by an OR gate and minimal path sets are of interest. The two cases are essentially the same since each is the Boolean dual of the other. The nature of the problem is illustrated by Table 7-1, which summarizes the results of an experiment made on a small fault tree headed by an AND gate and containing four branches. The baseline tree, the first line of Table 7-1, has 13 gates and 24 end events and contains 676 minimal cut sets. Resolution of these cut sets using the algorithm described in section 7.4.2 requires about 42 machine unit seconds (MUS) on an IBM 370-158-AP, about 16 minimal cut sets being reduced per MUS. In the experiment the baseline tree was incremented by adding two new end events at a time. The number of minimal cut sets in each successively larger tree and the number per MUS as additional end events are added to the tree are presented in Table 7-1.

The table shows that the number of minimal cut sets increases exponentially as end events are added to the tree while the number of cut sets resolved per MUS decreases exponentially. Consequently, the time required for solution of this set of trees increases at a rate that approaches a double exponential. Two techniques that allow minimal cut sets of trees with thousands of gates and end events to be resolved in reasonable time spans are

Figure 7-9
Sampling Gate Transformation

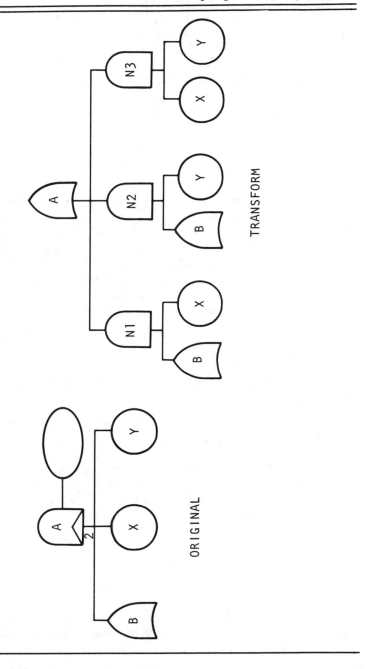

TRANSFORM

ORIGINAL

Table 7-1
Resolution Data

Tree	Number of Cut Sets	Number of Added Sets	Number Resolved Per MUS	Time in MUS Required
Baseline	676	0	16	42
1	936	260	16	59
2	1300	364	15	87
3	1800	500	12	150
4	2500	700	10	250
•	•	•	•	•
•	•	•	•	•
Six Additional end events	7848	2708	7	1000+

presented next. The first is bundling, which, although it can be implemented on a computer, allows resolution of fairly large trees to be accomplished rapidly without computers. The second, bit fiddling, is primarily suitable for computers. It is incorporated here because it illustrates a useful relationship between supersets and Boolean algebra manipulation.

7.8.1 Bundling

Bundling can be carried out for end events and gates. Each is discussed in turn.

7.8.1.1 End Event Bundling

The algorithm discussed in Section 7.4.2 is valid if those end events leading into an OR gate that are not duplicated elsewhere in the tree are *bundled*. That is, a substitution is made in which all end events are replaced by new ones before minimal cut sets are calculated. In Figure 7-10a, for example, the substitutions

$$Z1 = X4, \ X5, \ X6$$

$$Z2 = X7, \ X8, \ X9$$

can be made. The minimal cut sets of the new tree are $(Z1, X1)$ and $(Z2, X2, X3)$. Resolution into minimal cut sets of the original tree yields three two-

Figure 7-10a
Tree for Bundling Illustration

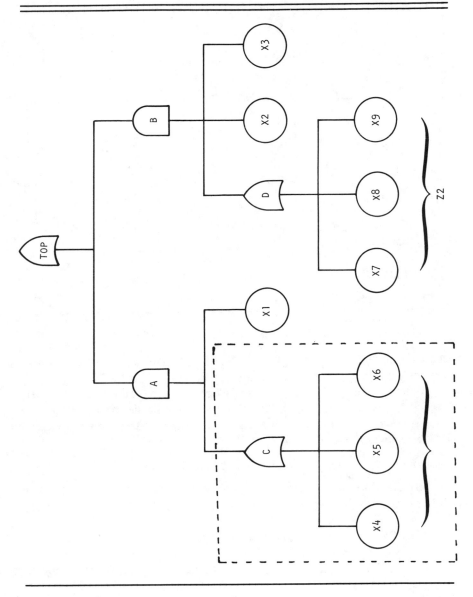

Figure 7-10b
Expansion of Figure 7-10a

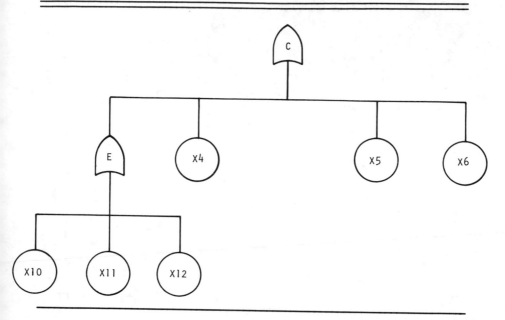

element cut sets, $(X1, X4)$, $(X1, X5)$, and $(X1, X6)$, and the three three-element cut sets $(X2, X3, X7)$, $(X2, X3, X8)$, and $(X2, X3, X9)$.

Bundling end events leading into OR gates can be carried out at higher levels if there are no intervening AND gates. Suppose the section of the tree inside the dashed rectangle of Figure 7-10a is expanded as shown in Figure 7-10b. A new substitution,

$$Z1' = X4, X5, X6, X10, X11, X12,$$

can be made. The minimal cut sets to TOP with this substitution are $(Z1', X1)$ and $(Z2, X3)$. Unbundling yields $(X1, X4)$, $(X1, X5)$, $(X1, X6)$, $(X1, X10)$, $(X1, X11)$, and $(X1, X12)$.

Bundling can also be used for inputs to AND gates, but time savings are not as great. Time savings by OR gate bundling can be several orders of magnitude. First, bundling yields a smaller tree to be resolved into minimal cut sets, thus avoiding the steep portion of the exponential curve (see Table 7-1) defining the time required for resolution for the number of minimal cut sets in a tree. Second, the time required for unbundling is a linear process. Consequently, there is a net transformation in large trees from resolution time that increases exponentially to resolution time that increases linearly.

7.8.1.2 Gate Bundling

A form of gate bundling is illustrated in Figure 7-11. The original tree can be transformed for minimal cut set calculation by eliminating all gates and end events above OR gate C. The ith minimal cut set leading to OR gate C is designated as C. Then the ith minimal cut set to TOP is, as shown in the figure, $C_i \cup X_1 \cup X_2 \cup X_3 \cup X_4 \cup D$ if X_1, X_2, X_3, X_4, or D are not duplicated elsewhere in the tree. If D, for example, is duplicated, then the ith minimal cut set leading to TOP is C_i because $C_i \cup X_1 \cup \ldots \cup D$ is a superset of C_i.

7.8.2 Bit Fiddling

While it is helpful, bundling may not be sufficient to permit minimal cut (or path) sets to be determined in large trees. A modification of the top-to-bottom algorithm which decreases resolution time by about 1.7 orders of magnitude is accomplished by transforming the end events into binary bits

Figure 7-11
AND Gate Bundling

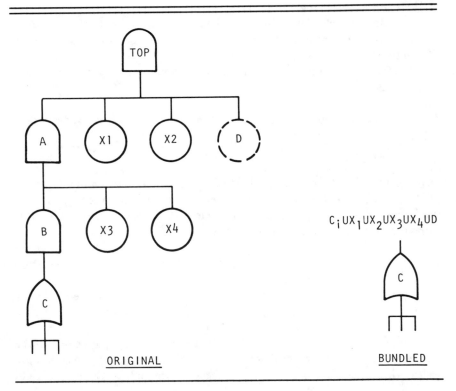

and carrying out the minimal cut set calculation by "bit fiddling." Bit fiddling refers to Boolean manipulations carried out in the computer's high-speed internal registers. This procedure not only decreases the time required for minimal cut set resolution but transforms the problem into one that is not dependent on core size. The combination of bundling and bit fiddling ensures that the minimal cut (or path) sets can be determined for any tree. The steps in the bit fiddling procedure are the following.

1. Rewrite the end event labels using integers $1, 2, \ldots, n$, where n is the number of end events in the entire tree.
2. Calculate the minimal cut (or path) sets to each of the gates at level 2 in the tree. This transforms the problem of finding minimal cut (or path) sets of one large tree to one of finding minimal cut (or path) sets of two or more smaller trees. In a typical test case, a variation of the trees shown in Figure 7-9, resolution to gate TOP (without bundling) was estimated to require several hours of computer time while resolution of all minimal cut sets of gates, A, B, C, and D required less than 20 MUS.
3. Transform each minimal cut (or path) set to a binary string such that the number that identifies a given element in a cut set is equivalent to a location in the string.
4. Combine the sets determined at level 2 to indicated cut sets at level 1. That is, the combinatorial process yields cut sets that may not be minimal.
5. Eliminate supersets, indicated cut sets containing supersets that are minimal.

Additional detail about these five steps is provided in the next few subsections.

7.8.2.1 Transformation to Binary Strings

Suppose that the first minimal cut set of subbranch A, written in integers, is $(2, 4, 6, 8)$. Suppose further that there are k cut sets in subbranch A, and that the largest of these k cut sets contains s elements. The cut set $(2, 4, 6, 8)$ is transformed into a binary string as shown in Figure 7-12 by

1. Establishing a $k \times s$ matrix for the A branch and initializing each location in this matrix to 0 (False).
2. Writing the minimal cut set $(2, 4, 6, 8)$ in the first row of the matrix by placing a 1 (True) in the second, fourth, sixth, and eighth positions.

The binary string shown in Figure 7-12 is, therefore, a unique analog of the cut set $(2, 4, 6, 8)$. The remaining $k - 1$ minimal cut sets leading to gate A are written in similar fashion, and this process is repeated for the minimal cut

Figure 7-12
Binary String Representation of Minimal Cut Set

POSITION	1	2	3	4	5	6	7	8	9	•	•	•	s
CONTENTS	0	1	0	1	0	1	0	1	0		•	•	0

sets leading to gates, *B*, *C*, and *D*. For illustrative purposes, let this be accomplished through matrixes of size $l \times t$, $m \times u$, and $n \times v$, respectively.

7.8.2.2 Combination at Top Level

Given four groups of minimal cut sets at level 2 in a tree, two steps are required to combine them into one set leading to the top. In the first, the union of all possible combinations is taken. As an illustration of this process, consider the result obtained by combining the first minimal cut set of each of the four second-level branches shown in Figure 7-13. The indicated cut set leading to the top gate is derived from the combination of the four cut sets shown in Figure 7-13 by the expression

$$A \cup B \cup C \cup D. \qquad (7.4)$$

If the widths of the matrixes, *s*, *t*, *u*, and *v*, are not too great, this can be programmed as a register-to-register operation in assembly language, the fastest possible executable instruction. The procedure is carried out by combining bits according to the rules for union given in Chapter 6. The result of carrying out the combination is described in Table 7-2. The first computer instruction, $A \cup B$, yields *A* since *B* comprises one bit in location 2.

Table 7-2
$A \cup B \cup C \cup D$

Location	1	2	3	4	5	6	7	8	9	10	11	...
A	0	1	0	1	0	1	0	1	0	0	0	...
B	0	1	0	0	0	0	0	0	0	0	0	...
*$A \cup B$	0	1	0	1	0	1	0	1	0	1	0	...
C	0	0	0	0	1	0	0	1	0	0	0	...
*$A \cup B \cup C$	0	1	0	1	1	1	0	1	0	0	0	...
D	0	0	0	0	0	0	0	1	0	1	0	...
*$A \cup B \cup C \cup D$	0	1	0	1	1	1	0	1	0	1	0	...

*Computer instruction

Figure 7-13
***Minimal Cut Sets for* A, B, C, *and D* Branches**

(2, 4, 6, 8)

	1	2	3	4	5	6	7	8	9	•	•	•	s
A(k=1)	0	1	0	1	0	1	0	1	0	•	•	•	0

(2)

	1	2	3	•	•	•	t
B(=1)	0	1	0	•	•	•	0

(5, 8)

	1	2	3	4	5	6	7	8	9	•	•	•	u
C(m=1)	0	0	0	0	1	0	0	1	0	•	•	•	0

(8, 10)

	1	2	3	4	5	6	7	8	9	10	11	•	•	•	v
D(n=1)	0	0	0	0	0	0	0	1	0	1	0	•	•	•	0

Adding *C* increases the size of the indicated cut set by one element, a bit in location 5; and adding *D* increases the size of the indicated cut set by adding a bit in location 10. The resultant indicated cut set in integer format is (2, 4, 5, 6, 8, 10). Combining binary strings in this fashion automatically eliminates duplicates and arranges the result in sequential order. For example, two binary sets whose integer formats are (4, 3, 6, 10) and (15, 10, 4, 7) are transformed to a binary string whose integer form is (3, 4, 6, 7, 10, 15). This characteristic aids in keeping machine time to a minimum.

7.8.2.3 Superset Elimination

Suppose the combination of the minimal cut sets leading to gates *A*, *B*, *C*, and *D* yields a matrix of indicated cut sets containing p rows and that the maximum number of elements in any one row is w. Let Figure 7-14 describe the first three indicated cut sets in the $p \times w$ matrix. In integer format the indicated cut sets are

Figure 7-14
Rows 1–3 of p $\times$ w *Matrix*

	1	2	3	4	5	6	7	8	9	10	11	• • •	w
p=1	0	1	0	1	1	1	0	1	0	1	0	• • •	0
p=2	0	1	0	0	1	0	0	1	0	0	0	• • •	0
p=3	0	0	1	0	0	0	1	0	1	0	0	• • •	0

Row	Cut Set
$p=1$	(2, 4, 5, 6, 8, 10)
$p=2$	(2, 5, 8)
$p=3$	(3, 7, 9)

The logic used to eliminate supersets is described in Table 7-3. Duplicates, trivial examples of supersets, are also removed by this logic. The first computer instruction shown in Table 7-3 carries out the logical instruction "a, EX-CLUSIVE OR b," written $a \oplus b$ in the table. This is done by combining bits according to the rules for exclusive OR gates described in Chapter 6. The results of the first step yield a binary string whose elements are in either a or b but not in both. The results of this logical operation are combined with set a and then with set b by the logical operator AND. These are written $(a \oplus b) \wedge a$ and $(a \oplus b) \wedge b$, respectively, and the rules of combination presented in Chapter 6 for AND gates are employed. The results of this operation yield two binary strings containing elements that are either a or b but not both, and

1. From the first AND operation are also in a, and
2. From the AND in the second operation are also in b.

The first intersection shown in Table 7-3 gives a nonzero result. This indicates that set b is not a superset of set a and must therefore be retained at this point pending further analysis. The results of the second intersection yield zeros in all positions of the binary string. Consequently, set a is a superset of set b, and set a is eliminated.

Table 7-3
Superset Elimination

Location	1	2	3	4	5	6	7	8	9	10	11	12	...
$a{:}p = 1$	0	1	0	1	1	1	0	1	0	1	0	0	...
$b{:}p = 2$	0	1	0	0	1	0	0	1	0	0	0	0	...
$c{:}p = 3$	0	0	1	0	0	0	1	0	1	0	0	0	...
$*a \oplus b$	0	0	0	1	0	1	0	0	0	1	0	0	...
$*(a \oplus b) \wedge a$	0	0	0	1	0	1	0	0	0	1	0	0	...
$*(a \oplus b) \wedge b$	0	0	0	0	0	0	0	0	0	0	0	0	...
$*b \oplus c$	0	1	1	0	1	0	1	1	1	0	0	0	...
$*(b \oplus c) \wedge b$	0	1	0	0	1	0	0	1	0	0	0	0	...
$*(b \oplus c) \wedge c$	0	0	1	0	0	0	1	0	1	0	0	0	...

*Computer instruction

This set of operations is replicated starting with the instruction $b \oplus c$. Table 7-3 shows that the intersections of subset b and c with the results of the operation $b \oplus c$ are nonzero. Therefore, b is not a superset of c, and c is not a superset of b.

7.9 IMPORTANCE OF CUT SETS AS A FUNCTION OF SIZE

If a tree contains no summation gates, it is generally true that minimal cut sets with smaller numbers of elements are more important than those with larger numbers. That is, it is generally true that a minimal cut set of size one is more important than a minimal cut set of size two; a minimal cut set of size two is more important than one of size three and so on. This follows from the fact that, in general, single events are more likely to occur than dual events; dual events are more likely to occur than triple events; and so on.

When, however, the tree contains summation, AND, and OR gates, the converse could be true. For the example in which the output event is "smog alert," inputs A, B, and C combining to cause an output is more likely to occur than exactly two inputs combining to cause an output, and two inputs combining is more likely to occur than one occurring in sufficient concentration to cause an output. Consequently, when the tree contains summation gates, minimal cut sets of size n are, in general, more likely to occur than minimal cut sets of size n-1, minimal cut sets of n-1 are more likely to occur than minimal cut sets of size n-2 and so forth.

7.10 COMMON CAUSE FAILURE ANALYSIS

Knowledge of the system's minimal cut sets permits common cause failure analyses to be conducted. *Common cause failure* (CCF) is defined as

Failure of multiple system components such that system failure results owing to one condition or cause. The single condition is referred to as a common cause event.

Concern for common cause failure has intensified in recent years in response to increased size and complexity of new technology. Larger jet aircraft, multi-state grids for distributing electricity, computer control of transportation systems, and genetic manipulation of microorganisms are but a few examples of such new technology. One procedure for analyzing common cause failures is to consider the set of events that can cause flow in all the members of at least one minimal cut set. In general, common cause failure candidates fall into the two major categories shown in Figure 7-15, engineering and operations. In turn, each may be subdivided into a set of generic causes such as the ones shown in the figure. A generic cause of failure is the result of a common cause event, such as excessive temperature. Excessive temperature resulting from a fire is not the same as excessive temperature caused by inadequate heat dissipation resulting from improper packaging or failures in the cooling system, even if one minimal cut set is being examined. An analysis of generic failure causes must, therefore, consider factors such as

1. Common location: An area of a building or vehicle where occurrence of a single generic cause could affect all components within that area.
2. Special condition: A characteristic of a basic event (an end event in a fault tree) shared by the elements of a minimal cut set.
3. Susceptibility: The tendency of a basic event to occur as a result of a generic cause of failure or special condition. The set of generic causes and special conditions to which a basic event is susceptible is known as a susceptibility set.

A positive indication in one of these areas may lend itself to recommendations for corrective action. For example,

1. Elements in a common location that are members of a susceptibility set can be rearranged so they are in more than one location or are afforded special protection.
2. Basic events that are susceptible to one generic cause can be made redundant, preferably in different form. For example, overspeed protection for a high-inertia rotating component could be provided by containment or by an electronic speed sensor that is redundant with a mechanical speed sensor or optical speed sensor.

In order to develop corrective action recommendations, it is sometimes helpful to classify common cause candidates in two categories, either of

Figure 7-15
Common Cause Failure Modes

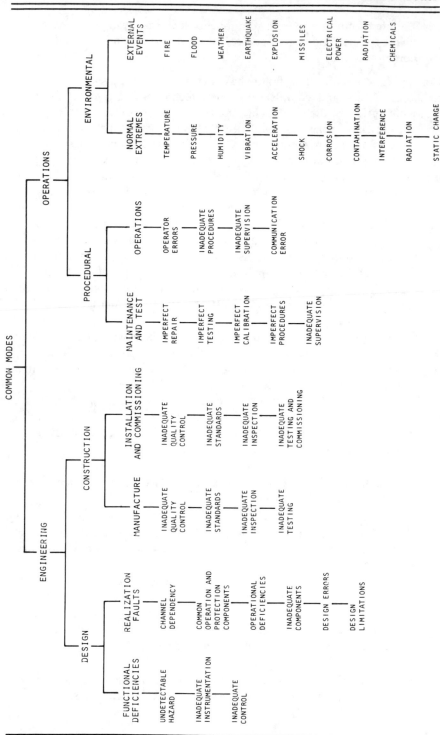

which may cause flow to occur in a minimal cut set: (1) environmental conditions and (2) conditions that couple all the basic end events in a given minimal cut set. In the former, all elements of the cut set must have a common location with regard to the environmental condition; common location may or may not be relevent in the latter case. As an example, consider a minimal cut set all of whose elements require periodic calibration or maintenance.

The following procedure can be used as a generic cause approach to common cause failure analysis.

1. Determine which minimal cut sets in a tree are of interest.
2. Determine which common modes of failure, using Figure 7-15 as an example, are applicable to the system.
3. For each basic end event, determine:
   a. Its susceptibility set
   b. Its sensitivity to each potential generic cause of failure in the basic event's susceptibility set
   c. The location of each component associated with the basic event
4. Determine whether the members of a cut set have a common location for each generic cause of failure.
5. Search the minimal cut sets for common cause candidates. It may be helpful to establish a minimum sensitivity for this step. (Computer aid is essential for this step if the system is complex.)
6. Group the common cause candidates by common cause event/common location doublets and special conditions. Study the system design, expected operational, and environmental conditions to identify possible mechanisms for each common cause event.
7. Form conclusions about the system's resistance to common cause failure and make recommendations for improvements in the system design to increase its resistance to common cause failure.

7.11 PRUNING THE TREE

Once all the minimal cut sets of a tree have been determined, it may be possible to prune those whose likelihood of occurrence is relatively low. If there are no summation gates, the minimal cut sets with the greatest number of elements are potential candidates for pruning. If there are only summation gates in the tree, the minimal cut sets of smallest size are potential candidates for pruning. If the tree contains summation and other gates, the important minimal cut sets can be distributed in a bimodal fashion, being among the minimal cut sets of smallest and largest sizes. One caveat should be noted: Pruning implies some degree of quantification, since only minimal cut sets that are unlikely to occur can be eliminated. The process of iterating an

Figure 7-16
Fault Tree Analysis Iteration

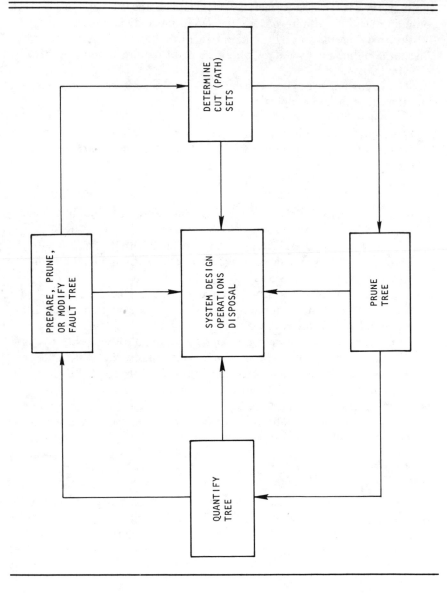

analysis by fault tree—drawing and pruning—incorporates quantification as shown in Figure 7-16. The relationship of this pruning to quantification is expanded further in Section 8.7.

The results of pruning yield a true fault tree since a tree is equal to the union of its minimal cut sets. An analogous situation exists for the dual tree when path sets are of concern. In that case the pruning is carried out to remove the minimal path sets containing the greatest number of elements. The kernel of the success tree is one in which the fewest number of elements must remain in operation for system success to be ensured.

Additional Reading

Barlow, R. E., J. B. Fussell, and N. D. Singpurwalla. *Reliability and Fault Tree Analysis*. Philadelphia: Society for Industrial and Applied Mathematics, 1975.

Barlow, R. E., and F. Proschan. Importance of system components and fault tree events. *Stochastic Processes and Their Applications* 3 (1975), pp. 1–21.

Chotterjee, P. *Fault Tree Analysis: Mincut Set Algorithms*. Operations Research Center Report ORC75-3. Berkeley: University of California, 1975.

Esary, J. D., and F. Proshan. Coherent structures with non-identical elements. *Technometrics* 5 (1963), pp. 191–209.

Fussell, J. B., E. B. Henry, and N. H. Marshall. *MOCUS: A Computer Program to Obtain Minimal Sets from Fault Trees*. UC-3. Idaho Falls, Idaho: Aerojet Nuclear Company, 1974.

Hogan, E. W. *Common-Mode/Common Cause Failure: A Review and Bibliography*. ORNL/NUREG, NSIC-148. Nuclear Safety Information Center, May 1979.

Malasky, S. W. and P. J. Tregarthen. *An Improvement in Cut and Path Set Determination*. Annual Reliability and Maintainability Symposium, 1980 Proceedings, New York: IEEE. ISSN 0149-144X. Pp 367–373.

Smith, A. M., and I. A. Watson. *Common Cause Failure: A Dilemma in Perspective*. Annual Reliability and Maintainability Symposium, 1980 Proceedings, New York: IEEE. ISSN 0149-144X. Pp 332–339.

Stockmeyer, L. J., and A. K. Chandra. Intrinsically difficult problems. *Scientific American*, May 1979, pp. 140–159.

Chapter 8

QUANTITATIVE FAULT TREE ANALYSIS

8.1 EVENT OCCURRENCE

It has been shown that system safety is primarily concerned with hazardous events and, in particular, with the likelihood of their occurrence. The binary operations (union and intersection) and the unary operation (complement), the fundamental mechanisms by which events may be mathematically manipulated, were introduced in Sections 3.2 and 3.3. Fault tree methodology discussed in Chapters 6 and 7 provides a set of descriptions by which such events may be logically combined. The occurrence of the top event in a tree may be described in terms of

1. Probability, a dimensionless number. Calculation of event occurrence in probabilistic terms is discussed in Section 8.2.
2. End event failure rate, which is expressed in terms of time. Transformation from time to probabilistic terms is approximated by multiplying the failure rate by a time factor, such as hazard duration or repair time. This is discussed in Section 8.5.

8.2 PROBABILITY CONSIDERATIONS

Consider a set of hazardous events, each event describable by a set of sample points in a sample space ω. The probability associated with these events is defined as follows:

> *Probability* is a function $P(\cdot)$ that assigns to every event E a nonnegative real number, denoted by $P(E)$, which is called the probability of the event E.

The domain of the probability function $P(\cdot)$ is the set of all events in ω. In order that $P(\cdot)$ be a probability function, $P(\cdot)$ must satisfy the following three postulates:

$$P(E) \geq 0 \qquad \text{for every event } E \qquad (8.1)$$
$$P(\omega) = 1 \qquad \text{for the certain event } \omega \qquad (8.2)$$
$$P(E \cup F) = P(E) + P(F) \quad \text{if } E \cap F = \Phi \qquad (8.3)$$

A set of hypotheses can be stated which satisfies these three postulates; those of significance to system saftey are discussed next.

8.2.1 Mutually Exclusive Events

It can be proven that Equation 8.3 is valid in its converse form. That is, $E \cap F = \Phi$ implies $P(E \cup F) = P(E) + P(F)$. In this expression the term $E \cap F = \Phi$ describes the situation in which events E and F are mutually exclusive. That is, two events are mutually exclusive if there are no sample points common to both. In Equation 8.3 this means that events E and F cannot both occur. Since by definition the safe and unsafe events S and $\bar{S}$ cannot both occur,

$$S \cap \bar{S} = \Phi \qquad (8.4)$$

In the general case, the probability of occurrence of a finite number of mutually exclusive events $E_1, E_2, \ldots, E_n$ is given by

$$P(E_1 \cup E_2 \cup \ldots \cup E_n) = P(E_1) + P(E_2) + \ldots + P(E_n) \qquad (8.5)$$

where

$$E_i \cap E_j = \Phi$$

The validity of Equation 8.5 can be proven by mathematical induction as follows:

1. $P(1)$ is true, since for this case Equation 8.5 becomes $P(E_1) = P(E_1)$.
2. Assume now that $P(n)$ is true and consider $P(n+1)$. Since $(E_1 \cup E_2 \cup \ldots \cup E_n)$ and E_{n+1} are mutually exclusive, $P(E_1 \cup E_2 \cup \ldots \cup E_{n+1}) = P(E_1 \cup E_2 \cup \ldots \cup E_n) + P(E_{n+1})$.
3. From this and the assumption that $P(n)$ is true, it follows that $P(E_1 \cup \ldots \cup E_{n+1}) = P(E_1) + \ldots + P(E_{n+1})$.
4. Consequently, $P(n)$ applies for any positive integer n.

Mutually exclusive events may be used as inputs to any of the logic gates in a fault tree, if a fault tree is an accurate model of the system. The occurrence of exactly two mutually exclusive events leading into an AND gate ensures that no output can occur because, by definition, only one of the two inputs can occur and both are required for flow through an AND gate. Therefore, the probability of flow occurring out of an AND gate with exactly two mutually exclusive events entering is identically equal to zero. Such a configuration is useful in a system if no output is a desirable characteristic.

8.2.2 Collectively Exhaustive Events

If the set of mutually exclusive events completely fills the universe such that one must occur, the events are said to be collectively exhaustive. Events $E_1, E_2, \ldots, E_n$ are said to be collectively exhaustive if

$$E_1 \cup E_2 \cup \ldots \cup E_n = \omega \qquad (8.6)$$

Collectively exhaustive events need not be mutually exclusive. For example, in a die toss the events $E_1, E_2, E_3, E_4, E_5,$ and E_6, corresponding to the numbers $1, 2, \ldots, 6$ on the faces of the die, are collectively exhaustive and mutually exclusive. The events E and F, throwing a number less than 4 and throwing a number greater than 2, respectively, are collectively exhaustive but not mutually exclusive. When events $E_1, E_2, \ldots, E_n$ are collectively exhaustive, Equation 8.5 equals one.

8.2.3 Events Not Mutually Exclusive

Suppose the output event of concern in a fault tree, E_o, is equal to some combination of events $E_1, E_2, \ldots, E_n$, but the system configuration is such that the E_n subevents are not mutually exclusive. In that case the probability of occurrence of the output event, that is, the probability that at least one input event will occur, is given by Boole's inequality,

$$P(E_o) \le P(E_1) + P(E_2) + \ldots + P(E_n) \qquad (8.7)$$

An equation rather than an inequality may, however, be desirable so that probability of occurrence of E_o can be determined more precisely.

When events are collectively exhaustive, Equation 8.7 can be transformed into an equality by taking into consideration the probability of intersections of the subevents and subtracting any which might be counted more than once. If, for example, exactly two events E and F are of concern, the probability that at least one will occur is given by

$$P(E \cup F) = P(E) + P(F) - P(E \cap F) \qquad (8.8)$$

Intuitive evidence of the validity of Equation 8.8 can be obtained by examining the Venn diagrams in Figure 8-1. In this figure, E and F are mutually exclusive, and there are no sample points common to both. In Figure 8-1(b) the crosshatched area is counted twice when the terms $P(E)$ and $P(F)$ are summed. Subtracting $E \cap F$ eliminates this duplication. Formal proof of Equation 8.8 is as follows:

Figure 8-1
Venn Diagrams

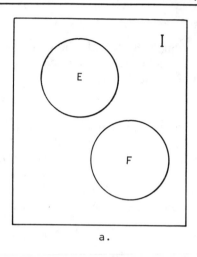

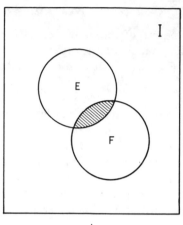

a.

b.

1. $E = E \cap \omega$
2. $\bar{F} \cup F = \omega$ Complementary law

Substituting step 1 into step 2 yields

3. $E = E \cap (\bar{F} \cup F)$

By means of the distributive law this becomes

4. $E = (E \cap \bar{F}) \cup (E \cap F)$
5. In a similar fashion, $F = (\bar{E} \cap F) \cup (E \cap F)$
6. Combining steps 4 and 5,
    $E \cup F = [(E \cap \bar{F}) \cup (E \cap F)] \cup [(\bar{E} \cap F) \cup (E \cap F)]$
    $= (\bar{E} \cap F) \cup (E \cap \bar{F}) \cup (\bar{E} \cap F)$
7. Now, since $(E \cap \bar{F})$ and $(E \cap F)$ are mutually exclusive,
    $P(E) = P(E \cap \bar{F}) + P(E \cap F)$
8. Similarly, $(\bar{E} \cap F)$ and $(E \cap F)$ are mutually exclusive, and
    $P(F) = P(\bar{E} \cap F) + P(E \cap F)$
9. Combining steps 6, 7, and 8,
    $P(E \cup F) = P(E \cap \bar{F}) + P(E \cap F) + P(\bar{E} \cap F)$

10. From step 7,
    $$P(E \cap \bar{F}) = P(E) - P(E \cap F),$$
    and from step 8,
    $$P(\bar{E} \cap F) = P(F) - P(E \cap F)$$
11. Substituting step 10 into step 9 yields Equation 8.8

If three events E, F, and G are collectively exhaustive, the probability that at least one will occur is given by

$$P(E \cup F \cup G) = P(E) + P(F) + P(G) - P(E \cap F) - \\ P(E \cap G) - P(F \cap G) + P(E \cap F \cap G) \qquad (8.9)$$

A Venn diagram of this case offers intuitive evidence similar to the two-event case. That is, if events E, F, and G have a common intersection, removal of the terms $(E \cap F)$, $(E \cap G)$, and $(F \cap G)$ from the expression for $P(E + F + G)$ removes the common area $E \cap F \cap G$ twice and it needs to be put back to retain the validity of the equality. Equations 8.8 and 8.9 can be illustrated by sets of sample points on a die. Let E be the set of numbers that are greater than 2 and let F be the set of numbers less than 5. Then, using sample points, $E \cap F = (3, 4)$. Therefore,

$$E \cup F = (3, 4, 5, 6) + (1, 2, 3, 4) - (3, 4)$$
$$= (1, 2, 3, 4, 5, 6)$$

Now let $G = 3$. Then $E \cap F = (3, 4)$, $E \cap G = 3$, $F \cap G = 3$ and $E \cap F \cap G = 3$. Then

$$E \cup F \cup G = (3, 4, 5, 6) + (1, 2, 3, 4) - (3, 4) - 3 - 3 + 3$$
$$= (1, 2, 3, 4, 5, 6)$$

8.3 INDEPENDENT AND DEPENDENT EVENTS

Consider the notions of dependence and independence. Informally, two events may be said to be independent if knowledge that one has occurred does not change the probability that the other will occur. As an example, consider two throws of a die. Knowledge of the first throw does not affect probability of any number occurring on the second die equal to $1/6$. Mathematically, independence of two events E and F can be expressed as

$$P(E \cap F) = P(E)P(F) \qquad (8.10)$$

Using this expression for calculating the probability of a single toss of a pair of dice, the probability of throwing $(1, 1)$ is

$$P(1 \cap 1) = P(1) \, P(1) = (1/6)(1/6) = 1/36$$

This result can also be obtained by counting sample points, as was done in Figure 3.1. That is, the universe consists of six sample points, and the selection of exactly one sample occurs with a probability of $1/6$. Extending the notion of independence to more than two events, events $E_1, E_2, \ldots, E_n$ are considered to be *pairwise independent* if, for any $i \neq j$,

$$P(E_i \cap E_j) = P(E_i) \, P(E_j) \qquad (8.11)$$

Pairwise independence for a set of n events need not imply that event $(E_i \cap E_j)$ and event E_k are independent of each other given $i \neq j \neq k$. A general criterion for independence, *mutual independence*, may be given as follows. Events $E_1, E_2, \ldots, E_n$ are mutually independent when, for any combination $1 \leq i < j < k < \ldots \leq n$,

$$P(E_i \cap E_j) = P(E_i) \, P(E_j)$$

$$P(E_i \cap E_j \cap E_k) = P(_i) \, P(E_j) P(E_k)$$

$$P(E_1 \cap E_2 \cap \ldots \cap E_n) = P(E_1) \, P(E_2) \ldots P(E_n) \qquad (8.12)$$

8.3.1. Independent Events Entering an AND Gate

Equation 8.12 describes the probability of flow from an AND gate when the inputs are mutually independent. If, for example, end events X, $Z1$, and $Z2$ in the fault tree in Figure 6-17 have probabilities of occurrence equal to 0.3, 0.2, and 0.1, respectively, then the probability of occurrence for C, the gate into which they flow, is given by

$$P(C) = (0.3) \, (0.2) \, (0.1) = 0.006$$

8.3.2 Independent Events Entering an OR Gate

If independent events $E_1, E_2, \ldots E_n$ flow into an OR gate, the probability of flow occurring out of the gate is given by

$$P(E_o) = 1 - \prod (1 - E_j) \qquad (8.13)$$

If, for example, end events $X1$, $X2$, and $Y5$ in Figure 6-17 have probabilities of occurrence equal to 0.3, 0.2, and 0.1, respectively, then the probability of occurrence of gate $(B\text{-}1\text{-}2B)$, is given by

$$P(B-1-2B) = 1 - (1-0.3) \, (1-0.2) \, (1-0.1)$$
$$= 1 - (0.7) \, (0.8) \, (0.9)$$

$$= 0.496$$

Therefore, the difference in probability of occurrence between the outputs of an AND and OR gate with identical inputs is about two orders of magnitude. Pruning (eliminating) the C branch in a subsequent iteration of the sample fault tree analysis might be reasonable.

8.3.3 Dependent Events

Two mutually exclusive events E and F are also independent if and only if $P(E)P(F)=0$. If this is true, it implies that either E or F has a probability equal to 0. Two events E and F are considered to be dependent when Equation 8.10 does not hold. That is,

$$P(E \cap F) \neq P(E)P(F) \tag{8.14}$$

Therefore, dependent events E and F should be used as inputs to the same logic gate only if the output of the gate is controlled by a modifier which expresses the nature of the dependence between these two events.

Some of the difficulties to be encountered in preparing a fault tree when the dependent, mutually exclusive qualities of events E and F are not explicitly stated may be illustrated by the following example. Consider two baseball teams, A and B, when both are playing on a specific afternoon. Let E be the event that A wins and F be the event that B wins. If A and B are not playing against each other, then the events E and F are independent, but not mutually exclusive. If the teams A and B are playing against each other, then the events E and F are mutually exclusive but not independent.

8.4 CONDITIONAL PROBABILITY

An extension to the notion of probability of events occurring in the future concerns the use of data already known about some of these events, or "conditional probability." *Conditional probability* may be defined as follows:

> Given events E_1 and E_2, the conditional probability of the event E_2 occurring given that the event E_1 has occurred is denoted by $P(E_2 \mid E_1)$. It is expressed as "the probability that E_2 will occur under the assumption that E_1 has occurred."

Mathematically, this may be written

$$P(E_2 \mid E_1) = \frac{P(E_1 \cap E_2)}{P(E_1)} \qquad \text{for } P(E_1) > 0 \tag{8.15}$$

where

$$P(E_2 \mid E_1) \text{ is undefined for } P(E_1) = 0$$

The expression for independence of two events, given by Equation 8.10, may be established more formally by using the notion of conditional probability. Suppose $P(E) > 0$, so that $P(F \mid E)$ is well defined. F is independent of E if the conditional probability of F, given that E has occurred, is equal to the unconditional probability of F. That is, E is independent of F if

$$P(F \mid E) = P(F) \qquad (8.16)$$

Suppose that both $P(F) > 0$ and $P(E) > 0$, so that $P(F \mid E)$ and $P(E \mid F)$ are well defined. Then from Equation 8.15,

$$P(E \cap F) = P(E)P(F \mid E)$$

From Equation 8.16,

$$P(E \cap F) = P(E)\ P(F)$$

By the commutative law (Equation 3.6), $P(E \cap F) = P(F \cap E)$. Therefore,

$$P(E \cap F) = P(F \cap E) = P(E)P(F) \qquad (8.17)$$

The following is a typical problem utilizing this notion. Suppose 100 safety valves are installed in a production line to prevent a hazard from occurring, and it is known that these safety valves come from a population in which one out of every 500 is defective. If one of the valves installed in the production line is found to be defective, what is the probability that another one of the remaining 99 is defective?

8.5 LAMBDA TAU APPROXIMATION

On occasion, flow out of an end event is expressed in terms of failures per unit time. As indicated earlier, failure rates can be multiplied by time so that flow out of each gate is expressed in dimensionless terms. Calculation of the probability of occurrence out of a gate whose inputs each have a known failure rate λ repair time or hazard duration τ may be approximated by the equations in Figure 8-2, provided:

1. λ_i and τ_i are reasonably constant
2. T/t, λ, and T are small
3. Failures occur independently

The procedure may be illustrated for an OR gate by examining the A branch of Figure 8.3. In this case,

$$\lambda_A = \lambda_x + \lambda_y$$
$$= 1 \times 10^{-5} + 1 \times 10^{-6} = 11 \times 10^{-6} \qquad (8.18)$$

Figure 8-2
Lambda Tau Expressions for Fault Tree Evaluation

Gate	τ's		2 INPUTS	3 INPUTS	n INPUTS
AND	UNEQUAL	λ	$\lambda_1\lambda_2(\tau_1+\tau_2)$	$\lambda_1\lambda_2\lambda_3(\tau_2\tau_3+\tau_1\tau_3+\tau_1\tau_2)$	$\lambda_1\lambda_2\cdots\lambda_n(\tau_2\tau_3\cdots\tau_n+\tau_1\tau_3\cdots\tau_n+\cdots+\tau_1\tau_2\cdots\tau_{n-1})$
	UNEQUAL	τ	$\dfrac{\tau_1\tau_2}{\tau_1+\tau_2}$	$\dfrac{\tau_1\tau_2\tau_3}{\tau_2\tau_3+\tau_1\tau_3+\tau_1\tau_2}$	$\dfrac{1}{\dfrac{1}{\tau_1}+\dfrac{1}{\tau_2}+\cdots+\dfrac{1}{\tau_n}}$
	EQUAL	λ	$2\lambda_1\lambda_2\tau$	$3\lambda_1\lambda_2\lambda_3\tau^2$	$n\lambda_1\lambda_2\cdots\lambda_n\tau^{n-1}$
	EQUAL	τ	$\dfrac{\tau}{2}$	$\dfrac{\tau}{3}$	$\dfrac{\tau}{n}$
OR	UNEQUAL	λ	$\lambda_1+\lambda_2$	$\lambda_1+\lambda_2+\lambda_3$	$\lambda_1+\lambda_2+\cdots+\lambda_n$
	UNEQUAL	τ	$\dfrac{\lambda_1\tau_1+\lambda_2\tau_2}{\lambda_1+\lambda_2}$	$\dfrac{\lambda_1\tau_1+\lambda_2\tau_2+\lambda_3\tau_3}{\lambda_1+\lambda_2+\lambda_3}$	$\dfrac{\lambda_1\tau_1+\lambda_2\tau_2+\cdots+\lambda_n\tau_n}{\lambda_1+\lambda_2+\cdots+\lambda_n}$
	EQUAL	λ	$\lambda_1+\lambda_2$	$\lambda_1+\lambda_2+\lambda_3$	$\lambda_1+\lambda_2+\cdots+\lambda_n$
	EQUAL	τ	τ	τ	τ

Figure 8-3
Sample Problem for Lamdba Tau Calculations

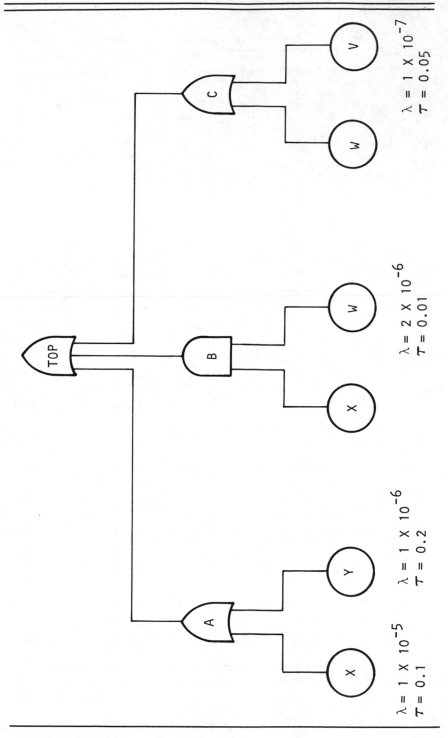

and

$$\Upsilon_A = (\lambda_x \Upsilon_x + \lambda_y \Upsilon_y)/\lambda_A$$
$$= [(1 \times 10^{-5})(0.1) + (1 \times 10^{-6})(0.2)]/11 \times 10^{-6}$$
$$= 1.2/11 = 0.11 \tag{8.19}$$

Consequently,

$$\lambda_A \Upsilon_A = (11 \times 10^{-6})(0.11) = 1.2 \times 10^{-6} \tag{8.20}$$

The procedure may be illustrated for an AND gate by examining the B branch of Figure 8.3. In this case,

$$\lambda_B = \lambda_x \lambda_w (\Upsilon_x + \Upsilon_w)$$
$$= [1 \times 10^{-5})(2 \times 10^{-6})(0.1 + 0.01)$$
$$= 2.2 \times 10^{-12} \tag{8.21}$$

and

$$\Upsilon_B = (\Upsilon_x \Upsilon_w)(\Upsilon_x + \Upsilon_w)$$
$$= (1 \times 10^{-3})(0.11) = 1 \times 10^{-4} \tag{8.22}$$

Consequently,

$$\lambda_B \Upsilon_B = (2.2 \times 10^{-12})(1 \times 10^{-4})$$
$$= 2.2 \times 10^{-16} \tag{8.23}$$

The probability of flow emanating from an AND gate with two inputs is much less than the probability expected from an OR gate with two inputs when their values for lambda and tau are similar.

8.6 ELIMINATION OF COMMON EVENTS

The probability of occurrence for OR gate TOP cannot be obtained by direct combination of the outputs of gates, A, B, and C in Figure 8-3 because there are common events in the tree as well as events X and W. In terms of minimal cut sets, it can be seen that the single cut set in the B branch, (X, W), is a superset of the minimal cut set X in the A branch and the minimal cut set W in the B branch. When the duplicates in a tree are known, they can be eliminated in the lambda-tau approximation by

1. Making no change when they meet at an OR gate
2. Setting all but one of the common events equal to 1 when they meet at an AND gate

Expansion by Boolean algebra will automatically eliminate duplicate end events. In Figure 8-3, for example,

$$TOP = A \cap B \cup C$$

$$= (X \cup Y) \cup (X \cap W) \cup (W \cup V) \qquad (8.24)$$

Let $(X \cap W) = M$ and $(W \cup V) = N$. Substituting these values in Equation 8.24 yields

$$TOP = (X \cup Y) \cup M \cup N$$

Using the distributive law, this can be written

$$TOP = Y \cup (X \cup M) \cup N$$

Substituting for M yields

$$TOP = Y \cup [X \cup (X \cap W)] \cup N$$

Using the absorption law and substituting for N, this becomes

$$TOP = Y \cup X \cup W \cup V \qquad (8.25)$$

Equation 8.25 indicates that TOP can occur if any one or more of the end events X, Y, W, or Z occur. But these four end events are the minimal cut sets of the tree shown in Figure 8-3. Therefore, a tree composed exclusively of OR gate TOP and the end event X, Y, W, and Z may be used for calculating probabilities of occurrence of the top event by lambda-tau or some other method. Since a tree can be formed from the union of all its minimal cut sets, this process yields a valid model. However, a tree formed in this fashion, an OR gate with four end events, although useful as an equivalent tree for minimal cut set evaluation, is not equivalent as a system model for determining the minimal cut sets of the dual tree.

A more accurate calculation for safety can be made using the expressions in Table 8-1 when failure rates are known for each end event in a fault tree. These equations describe three differential and two integral calculations for each end event, minimal cut set, and the system for each time selected. The quantities calculated are:

Q: The probability that an end event (or cut set, or system) is in a failed condition at $q(t)$.

W: The expected component (cut set or system) failure rate per hour at $W(t)$.

L: The probability that the component (cut set, or system) will fail in the interval t to $t + dt$, given that it is functioning at t. (For an end event, this is equal to lambda.)

Table 8-1
Expressions for Safety Calculations

Q

Components	$q(t) = 1 - \exp\left[\int_0^t -\lambda(t')dt'\right] = 1 - e^{-\lambda t}$
Cut sets	$Q_k(t) = \prod_{j=1}^{m} q_i(t)$
System upper bound	$Q_s(t) = 1 - \prod_{i=1}^{N_c} [1 - \check{Q}_i(t)]$

W

Components	$W(t) = \lambda(t)[1 - q(t)] = \lambda \cdot e^{-\lambda t}$
Cut sets	$W_k(t) = \sum_{j=1}^{m} W_j(t) \prod_{\substack{l=1 \\ l \neq j}}^{m} q_l(t)$
System upper bound	$W_s \leq \sum_{1=i}^{N_c} W_i(t)$

L

Components	$L = \lambda$
Cut sets	$\Lambda_k(t) = \dfrac{\sum_{j=1}^{n} W_j(t) \prod_{\substack{l=1 \\ l \neq j}}^{n} q_i(t)}{1 - Q_k(t)} = \dfrac{W_{ki}(t)}{1 - Q_k(t)}$
System upper bound	$\Lambda_s(t) \leq \dfrac{\sum_{i=1}^{N_c} W_{si}(t)}{\prod_{i=1}^{N} [1 - Q_s(t)]} = \leq \dfrac{W_s(t)}{1 - Q_s(t)}$

WSUM

Components	$\int_{t_1}^{t} w(t')dt' = e^{-\lambda t_1} - e^{-\lambda t}$
Cut sets	$\int_{t_1}^{t} w_k(t')dt'$
System upper bound	$\int_{t_1}^{t} \left[\sum_{i=1}^{N_c} W_{ki}(t') \right] dt' \leq \int_0^t W_o(t')dt'$

Table 8-1
(Continued)

FSUM

Components	$1 - \exp[-\lambda(t - t_1)]$
Cut sets	$1 - \exp\left[-\int_{t_1}^{t_2} \check{\Lambda}(t)dt\right]$
System upper bound	$1 - \exp\left\{-\int_{t_1}^{t_2}\left[\dfrac{\sum\limits_{i=1}^{N_c} \check{W}_{ki}(t)}{\prod\limits_{i=1}^{N_c}[1 - \check{Q}_{ki}(t)]}\right]dt\right\} \leq$
	$1 - \exp\left[\int_{t_1}^{t_2} \Lambda(t)dt\right]$

WSUM: The expected number of failures of the end event (cut set or system) to time t. That is,

$$\int_0^t w(t')dt'$$

FSUM: The probability of one or more failures to time t. That is,

$$1 - \exp(-\lambda t)$$

This same information may be used to calculate the relative importance of the cut sets and events. The expressions for these calculations are shown in Table 8-2.

Table 8-2
Importance Calculations

Cut set importance	$I_k(t) = \dfrac{\check{Q}_k(t)}{Q_s(t)}$
Component Importance	$i_x(t) = \sum\limits_{i=1}^{n} \dfrac{Q_{kj}(t)}{Q_s(t)}$
	where
	$j =$ the jth cut set oj the n cut sets containing x

8.7 END EVENT IMPORTANCE

End event importance can be considered in terms of three factors:

1. The probability of occurrence of the end event. However, if an end event is a member of an *n*-element minimal cut set, the probability of joint occurrence of all members of the minimal cut set is of concern.
2. The probability that the top event occurs in the tree given the occurrence of the end event or minimal cut set.
3. The significance to the system and its surrounding environment associated with occurrence of the top event.

Most often, the criteria of concern is some combination of probability of occurrence and resultant consequence. A scheme for combining these two is suggested by Figure 8-4 in which the cross hatched section is presumed to represent the area of concern. In this figure, for example, preventive measures are not pursued for Class I hazards determined to be unlikely to occur, and, conversely, problems of little consequence are not resolved even though their probability of occurrence is assessed to be high.

The planar character of Figure 8-4 may be changed to three dimensions by adding weight to certain safety concerns or by incorporating nonsafety concerns, such as cost and schedule. Suppose the left-hand edge of the crosshatched area with coordinates (unlikely, I) in Figure 8-4 is made up of four sample points, each reflecting a unique failure mode. While each of these failure modes has about the same likelihood of occurrence, and each results in a Class I consequence, their effect can, nevertheless, vary considerably. A range of one to hundreds of deaths is not distinguishable in this figure, for each is a Class I hazard. If, however, a *z* coordinate is added in the vertical direction, the relative height above the *x-y* plane in Figure 8-4 would indicate the comparative importance of the four sample points. In like fashion, differing failure modes having approximately equal likelihood of occurrence and hazard classification may vary significantly in the importance of nonsafety considerations such as repair costs should the hazard occur, costs required to significantly lower the likelihood of occurrence, or effect on schedule, in terms of costs and/or time. Such nonsafety factors may also be expressed as height along a *z* axis above the *x-y* plane of Figure 8-4.

8.8 COMBINING THE FTA AND HMEA

The obvious difference between the fault tree analysis (FTA) and the hazard mode and effects analysis (HMEA) is one of format, but this does not prevent the merging of an FTA and HMEA. There are, however, differences in content which do prevent merging the two. The primary difference is

Figure 8-4
Importance Considerations

HAZARD CLASS	PROBABILITY OF OCCURRENCE			
	UNLIKELY	LOW	MODERATE	HIGH
I		▨	▨	▨
II			▨	▨
III				▨
IV				▨

caused by the fact that the top event in a fault tree, while it does not restrict the nature of the events that may cause it to occur, is itself specific and therefore somewhat restricted, while top events developed in an HMEA from hypothesized failures of detail parts are unrestricted. In order to merge the two, it is necessary for the HMEA to be made more like a fault tree or for the fault tree to be made more like an HMEA. Of the two, the latter is the more desirable choice because

1. Fault trees are more general in that there are no restrictions on the nature of the elements beneath the top event.
2. Computer technology makes the fault tree easier to manipulate.

One procedure for merging the two begins with a selection of the set of minimal cut sets in the fault tree that are most important. If probabilities (or failure rates) are not known or estimable for all the end events in the tree, a sensitivity analysis can be performed. In this analysis the probability of occurrence of the top event is determined as a function of variation in probability of occurrence of the end events. The result will indicate which end events strongly affect the failure rate of the system when their failure rates are changed by a small amount. The tree may then be redrawn using only the minimal cut sets deemed to be most significant. That is, only those cut sets whose probability of occurrence is high or which can strongly alter the system failure rate through small changes in end event failure rates are used.

Next a criticality index is established for the top event in the fault tree for each minimal cut set selected as being important. Like the criticality index in the bottom-to-top hazard mode, effects, and criticality analysis, the criticality index associated with the top event in the fault tree is some combination, the product, perhaps, of probability of occurrence of the top event and resultant consequence. If, for example, the top event is "accident occurs," the occurrence of each minimal cut set selected can be reviewed to assess the precise implications associated with occurrence of the top event. Does occurrence of the top event imply multiple fatality or minor injury, significant or minor loss of property, profound or little effect on production schedules or sales? As in the case of the criticality index developed for a bottom-to-top hazard analysis, the result obtained by multiplying probability of occurrence and a number, say 1 to 4, associated with a significance assigned to the nature of the event can be modified by weighting factors. Suppose weighting factors are permitted to range from 1 to 20. Some typical values could be,

1. Fatality: 20.
2. Nonfatal injury: 1–20, depending on the nature of the injury and the number of people affected, with a maximum value of 20.
3. Schedule loss: n points per unit of time, with a maximum value of 20.
4. Costs per occurrence: m points per d dollars, with a maximum value of 20.

Areas may then be selected for complete analysis by HMEA technology through selection of a value for K in the expression

$$P_i \, C_i \, W_i > K \text{ for } (0 \leq K \leq 100) \tag{8.26}$$

where

$P_i =$ the probability of occurrence of the ith minimal cut set
$C_i =$ the criticality index (1–4) associated with occurrence of the ith minimal cut set
$W_i =$ the weighting factor (1–20) assigned to the top event for occurrence of the ith minimal cut set.

The process of combining the FTA and HMEA is iterative. That is, the material developed by the HMEA for the areas selected in the fault tree may be used to update the fault tree.

8.9 FAILURE RATE ASSESSMENT

The accuracy of the failure rate (or probability of occurrence) determined for the system depends upon the accuracy with which the failure rates (or probabilities of occurrence) are known for the end events. Ideally, this information is based upon controlled test data obtained from production samples. In practice, particularly when the fault tree model is developed to predict rather than measure system behavior, failure rate data is often derived by mathematical modeling or by extrapolation of data collected from similar equipment. Help is available for estimating failure rates from various organizations formed for the purpose of gathering test and usage data. Some organizations are highly specialized, relating to a specific industry, such as agriculture, forestry, or medicine. Two more generally oriented organizations are discussed next.

8.9.1 Government-Industry Data Exchange Program

The United States government funds and manages a Government-Industry Data Exchange Program (GIDEP) for exchanging technical data used in research, design, development, production, and operations of systems. Almost all U.S. government agencies, hundreds of industrial organizations, and the Canadian Department of Defense participate in the program. Electronic part and component data are also exchanged with the European EXACT program. Since participation is mandatory for many Defense Department and NASA activities, new data routinely flow into the four data banks maintained:

An Engineering Data Bank (EDB) that contains engineering evaluation and qualification test reports, nonstandard parts justification data, parts and materials specifications, manufacturing processes, failure analysis data, and other related engineering data on parts, components, materials, and processes. This bank also includes a section of reports on specific engineering methodology and techniques, air and water pollution reports, alternative energy sources, and other subjects.
A Reliability-Maintainability Data Bank (RMDB) that contains failure rate/mode and replacement rate data on parts and components based on field performance information and/or reliability demonstration tests of equipment, subsystems, and systems. This data bank also contains

reports on theories, methods, techniques, and procedures related to reliability and maintainability practices.

A Metrology Data Bank (MDB) that contains related metrology engineering data on test systems, calibration systems, and measurement technology and test equipment calibration procedures. This bank has been designated as a data repository for the National Bureau of Standards (NBS) metrology-related data. It also provides a Metrology Information Service (MIS) for participants.

A Failure Experience Data Bank (FEDB) that contains failure information generated when significant problems are identified on parts, components, processes, fluids, materials, or safety and fire hazards.

Any activity that uses or generates data suitable for inclusion in the GIDEP data banks may be eligible to participate. Consideration is being given to the notion that GIDEP should be made the focal point for future rate information on hardware used in radiation environments.

8.9.2 Reliability Analysis Center

The Defense Logistics Agency sponsors a Reliability Analysis Center (RAC) that is managed by the Rome Air Development Center (RADC) and operated at RADC by IIT Research Institute (IIRTI). RAC is charged with the collection, analysis, and dissemination of failure rate data for electronic and nonelectronic components. Data for the parts are grouped into about twelve environments, varying from benign laboratory conditions to severe missile launch conditions. These cover ranges from nearly no environmental stress with optimum engineering and maintenance to severe conditions of noise, vibration, and other environments. The failure rates are further refined to reflect other stresses such as mechanical and maintenance considerations. Much of the data is maintained on computers which are programmed to look up failure rate data and to combine the failure rates for individual components into rates for subassemblies and systems. Combination of component failure rates into system failure rates requires that block diagrams and redundancy information be provided.

8.10 GENERIC FAILURE RATES

Failure rates for any condition that can be expressed in terms of those established for ideal laboratory conditions are known as generic failure rates. A *generic failure rate*, for a component or system, may be defined as

> The inherent number of failures per unit of time, cycles, or trials that occur to the equipment under laboratory conditions.

Similarly, *generic life expectancy* may be defined as

> That point in time (or number of cycles or trials) when the failure rate of the equipment, operated under laboratory conditions, increases beyond some specified value.

Weighting factors can be applied to generic failure rates and life expectancies to reflect environments more severe than the laboratory. Generic failure rates are relatively invariant for any given item, but generic life expectancy will vary in accordance with the safety requirements established for the system. Some increase in generic life expectancy can be expected if the amount of maintenance performed on the system is increased. Eventually, however, life expectancy is limited by the amount of time or number of cycles, which causes wearing out.

8.11 ESTIMATION OF HUMAN FAILURE RATES

Human failure is considered to occur when:

1. An individual fails to perform a task or some portion of a task.
2. A task is performed incorrectly.
3. A step is introduced into the process which should not have been included.
4. A step is conducted out of sequence.
5. A task is not completed within an allocated time span.

The ability to estimate the frequency with which hazardous events can be caused by human error is more difficult than estimating the expected frequency of hardware failures. The fundamental problem which prevents estimating human failure rates is the inability to quantify the reliability of human performance in the mathematical language of the systems analyst. For example;

1. The distinction between failure and nonfailure is easier to establish for equipment than it is for personnel. In part, this is due to the fact that human failure, presuming that medical conditions are not involved, is often equated with human error and there may be some reluctance to formalize the commission of an error. No such reluctance exists for reporting equipment failures.
2. Equipment failures, especially at lower hardware levels, depend only upon the stresses imposed on the equipment, including those established by the ambient environment, rather than upon the form of the system or the nature of its output. Consequently, generic failure rates established

for a given time interval remain relatively invariant from system to system. However, an estimate of the number of human errors to be expected in a certain man-machine interfacing system may not be valid for other, even strongly similar systems.

3. The effects of maintenance, both preventive and corrective, are more difficult to predict for human systems than for hardware systems. In this regard, it should be noted that sleep can be classified as routine maintenance for human systems. The limitations imposed in commercial aviation upon the number of hours that a flight crew are permitted to work is one instance in which predicted human error is related to system operating time.

One method for performing a reliability analysis of human systems, in use since 1961, is the Technique for Human Error Rate Prediction (THERP) developed by Swain and Rook. THERP is intended to quantify predictions of degradations that result from human errors in an environment influenced by problems of equipment unreliability, operational procedures, and any other characteristic of the system that influences human behavior. The interface between THERP and system safety is established by developing those sample points that are common to the domains of human error and of hazardous events. The process of developing those sample points that are synonymous with human error is an iterative procedure consisting of five steps:

1. Determine which system failure modes are to be evaluated.
2. Identify the significant human operations that are required and their relationships to system operation and system output.
3. Estimate error rates for each human operation or group of operations that are pertinent to the evaluation.
4. Determine the effect of human errors on the system and its outputs.
5. Modify system inputs or the character of the system itself so as to reduce the failure rate.

These five steps may be iterated, although not necessarily in the order listed, until system degradation resulting from human error is at an acceptable level. In applying THERP, the operational situation is usually modeled as a probability tree where the branches of the tree represent a diagrammatic task analysis that illustrates the sequence and flow of task behaviors and other situational relationships. Each branch may be assigned a discrete probability of occurrence or nonoccurrence. Probabilistic relationships, such as dependence versus independence of events and the effects of stress, climate, motivation, equipment malfunctions, and various operational contingencies, may also be included in the construction of the probability tree. Once the tree has been generated, exercising the model follows the prediction techniques established for fault trees.

Additional Reading

Albert, A. A. *Modern Higher Algebra*. Chicago: University of Chicago Press, 1947.

Carmichael, R. D. *Introduction to the Theory of Groups of Finite Order*. Lexington, Mass.: Ginn, 1937.

Government-Industry Data Exchange Program. GIDEP Operations Center, Corona, Calif. 91720.

Halmos, B. R. *Introduction to Hilbert Space*. New York: Chelsea Publishing, 1951.

Hohn, F. E. *Applied Boolean Algebra*. New York: Macmillan, 1960.

Nonelectronic Parts Reliability Data. NPRD-1. Reliability Analysis Center, Rome Air Development Center, Griffiths AFB, N.Y. 13441.

Parzen, E. *Modern Probability Theory and Its Applications*. New York: Wiley, 1960.

Reliability Prediction of Electronic Equipment. Military Standardization Handbook 217. Washington, D.C.: U.S. Government Printing Office.

Swain, A. B. *THERP*, Proceedings of the Symposium on the Quantification of Human Performance. Albuquerque: University of New Mexico, 17–19 August 1964.

U.S. Naval Fleet Missile Systems and Analysis and Evaluation Group. *Failure Rate Data Handbook*. Corona, Calif.: U.S. Navy, Oct. 1976.

Chapter 9

UNCERTAINTIES IN SAFETY MEASUREMENTS

9.1 BASES FOR UNCERTAINTY

Quantitative assessments of hazardous occurrences strongly influence programs undertaken to ensure a safe product or activity. That is, great effort is readily undertaken to eliminate hazards assessed to be Classes I and II, and fewer efforts can be expected for those assessed as III and IV. Consequently, an inaccurate assessment can cause resources to be used inefficiently or allow serious hazards to remain in a system. Consideration needs to be given, therefore, to those factors that could cause such assessments to be erroneous. Specifically, unreliability of numerical safety estimates is mainly attributable to three factors:

1. An accurate system safety estimate implies completion of an optimization process. However, as indicated in Section 2.9, attempts to optimize invariably reduce to suboptimization, mainly as a consequence of the difficulty of considering all possible alternatives and all possible outcomes.
2. The suboptimization process requires value considerations to be used in the estimating process (see Section 2.10). By their very nature, value judgments permit uncertainty to be introduced.
3. The raw data used in estimating must contain some uncertainty. It is this aspect of uncertainty that is discussed in this chapter.

It is a matter of fact that numerical data used in solving problems in everyday life are not exact. A simple measurement like length is determined to some value plus or minus some uncertainty. Errors in defining a safety level through use of such data, therefore, result in some uncertainty. In addition, all applicable data may not be available for the estimating process. A hazardous event, for example, may have been improperly reported or not reported at all. Safety levels may, therefore, be based in part upon inference, requiring application of inductive reasoning. In this process, one goes from data available about the particular to general data, with results that can incorporate some of the inaccuracies in the raw data.

Although inductive reasoning can lead to difficulties in describing the general case, it is possible to be precise about results obtained by this process in terms of the risks involved in making such inferences. In this context, "risk" is used in the speculative sense, the odds that may be given to the accuracy of the inference, that is, the probability of its accuracy. Inferring safety levels by inductive reasoning is done through the use of distributions. This permits inferences to be made about the universe of systems under consideration, based upon particular samples obtained from this universe. More formally, probability distributions are used in testing statistical hypotheses and assessing the significance of experimental results.

9.2 RANDOM PHENOMENA AND EVENTS

Events were described in Section 2.2 as a collection of defined sample points. Input and output events, whether they are one or more sample points, are the building blocks for the construction of fault trees; probability theory is the binder with which these building blocks are connected. The procedure by which the connections are made depends in part on whether the events used in the construction are random. Specifically, estimating the occurrence of a hazardous event, such as the number of automobile accidents that can be expected to occur on a given highway during some interval of time, uses the notions of random (chance) phenomena and random events.

> A *random phenomenon* is an empirical occurrence whose observed outcomes under similar circumstances are not always identical, but whose observed outcomes do have statistical (rather than deterministic) regularity.

This definition implies that numbers exist between 0 and 1 that represent the relative frequency with which the different possible outcomes may be observed in a series of observations of independent occurrences of the phenomenon. Closely related is the notion of random event and the probability of the occurrence of a random event.

> A *random event* is one whose relative frequency of occurrences, as observed in a sequence of observations of randomly selected situations in which the event can occur, approaches a limiting value as $n \to \infty$.

The limiting value of this relative frequency is the probability of occurrence of the random event. Using automobile accidents as an example, it can be perceived that the occurrence of any accident at some location depends upon a number of variable circumstances, each of which controls the arrival of two vehicles at a precise point in time and location. A decision to have (or not to have) a second cup of coffee, selection of an alternate route, an unusual noise causing a driver to slow down a bit, or the need to stop for a pedestrian or stray pet can alter the timing and permit (or prevent) the accident rendezvous from occurring. Nevertheless, data collected may indicate a reasonably constant and, therefore, predictable number of accidents for a particular location as a function of weather, time of day, and day of the week. Consequently, odds may be established a priori that indicate the probability that an individual will have an accident at that location.

If this probability remains relatively constant for some set of weather and time conditions, it may be implied that what happens to a motorist at this location under those conditions is a random phenomenon and that the occurrence of an accident is a random event. Expressing this relationship more generally, a numerical-valued random phenomenon can be described by stating its probability function. However, the probability function $P(\cdot)$ is a function of sets and, consequently, may be unwieldy to treat in an analytical fashion. A more desirable circumstance is to have $P(\cdot)$ represented by a function of points, that is, let it be a function of real numbers x. It may be shown that for any given numerical-valued random phenomenon, there exists a point function, called the (cumulative) distribution function, which is sufficient for determining the probability function in the sense that the probability function may be reconstructed from the distribution function.

The distribution function $F(\cdot)$ of a numerical-valued random phenomenon is defined as having as its value at any real number x the probability that an observed value of the random phenomenon will be less than or equal to the number x. Symbolically, for any real number x,

$$F(x) = P(\text{real numbers } x':x' \leq x) \tag{9.1}$$

It is equivalent to state that $F(x)$ is a distribution function if

$$0 \leq F(x) \leq 1$$
$$F(-\infty) = 0; \; F(\infty) = 1$$
$F(x)$ is a nondecreasing function
$F(x)$ is continuous to the right $\tag{9.2}$

9.3 PROBABILITY MASS AND DENSITY FUNCTIONS

In most cases of concern to system safety that deal with random phenomena, the distribution functions describing these phenomena have probability functions that may be specified by a probability mass or density function. Since these two probability functions are point functions, they will suffice for describing most random phenomena of concern.

If the probability function is specified by a probability mass function p, then the distribution function $F(x)$, for any real number x, is given by

$$F(x) = \Sigma \, p(x') \text{ for points } x' \leq x \text{ such that } p(x') > 0 \qquad (9.3)$$

If the probability function is specified by a probability density (frequency) function $f(x)$, then the corresponding distribution function $F(x)$, for any real number x, is given by

$$F(x) = \int_{-\infty}^{x} f(x') \, dx' \qquad (9.4)$$

The term "mass function" and "density function" came into use in this domain through the analogy that exists for these terms in the physical world. Suppose material of unit mass is distributed over a line segment, such that the amount of distributed material over any set of N real numbers on the line segment is equal to $P(N)$. If there is a positive amount of material at the point x, the distribution of material possesses a mass at the point x which may be denoted by $p(x)$, as shown in Figure 9.1. If, for any interval of the line segment ℓ that contains the point x (with ℓ being a sufficiently small number) the distribution of material has a density at the point x which may be denoted by $f(x)$. The weight of the material attached to the interval ℓ is equal to $\ell f(x)$, as illustrated in Figure 9-2.

The graph of the mass distribution function shown in Figure 9-1 may be expressed as

$$
\begin{aligned}
f(x) &= 1/40 & &\text{for } x = 0, 4, 7, 8, 12, 13, 15, 16, 17, 18 \\
&= 1/16 & &\text{for } x = 2, 10, 14, 19 \\
&= 1/20 & &\text{for } x = 1, 3, 6, 9, 11 \\
&= 1/8 & &\text{for } x = 5, 20 \\
&= 0 \text{ except} & &\text{for } x = 0, 1, \ldots, 20 \qquad (9.5)
\end{aligned}
$$

The algebraic expression for the probability density function shown in Figure 9-2 can be written as

$$
\begin{aligned}
f(x) &= 0 & &\text{for } x < 0 \text{ and } 10.125 \leq x \\
&= 0.05 \,(x+1) & &\text{for } 0 \leq x < 1.0 \\
&= 0.1 & &\text{for } 1.0 \leq x < 2.0 \text{ and } 6.0 \leq x < 10.125 \\
&= 0.1 \,(x-1) & &\text{for } 2.0 \leq x < 3.0 \\
&= 0.2(4-x) & &\text{for } 3.0 \leq x < 3.5 \\
&= 0.025(7.5-x) & &\text{for } 3.5 \leq x < 5.5 \\
&= 0.1(x-5.0) & &\text{for } 5.5 \leq x < 6.0 \qquad (9.6)
\end{aligned}
$$

Figure 9-1
Graph of Probability Mass Function

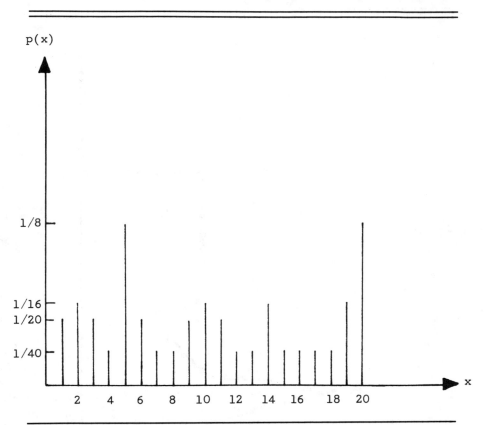

A probability function $P(\cdot)$ always has a probability density function and a probability mass function. In order to specify a probability function by either, it is necessary that the other vanish identically. More often it is the probability density function that is selected to describe the random phenomena of concern to system safety.

9.4 DEFINITION OF PROBABILITY DENSITY FUNCTION

There exists a function $f(\cdot)$ for many probability functions from which $P(S)$ can be obtained for any event S by the expression

$$P(S)=\int_{s} f(x)dx \qquad (9.7)$$

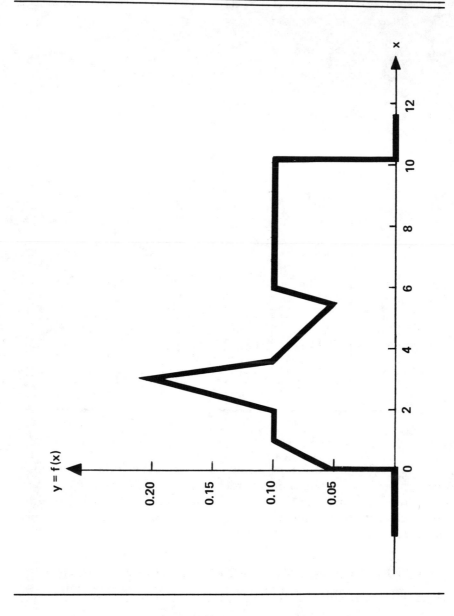

Figure 9-2
Graph of Probability Density Function

The function $f(\cdot)$ is a probability density function if

$$\int_{-\infty}^{\infty} f(x)dx = 1 \qquad (9.8)$$

and

$$f(x) \geq 0 \text{ for all } x \text{ in } S \qquad (9.9)$$

The quantity $f(x)dx$ is known as the "probability density element," or "frequency element," at the point x and is a measurement of the probability of the interval $(x, x + dx)$. The probability in any interval between a and b,

$$a < x < b, \ a \leq x \leq b, \ a < x \leq b \text{ or } a \leq x < b,$$

is given by

$$\int_{a}^{b} f(x)dx \qquad (9.10)$$

if the distribution is continuous in that interval.

9.5 GENERAL DISTRIBUTION FOR FAILURE OCCURRENCE

A general distribution for describing the occurrence of failures of large groups as a function of time is presented in Figure 9-3. The figure has the same form whether it is considered a frequency function, in which case the number of failures is used as the ordinate, or it is considered a probability density function, when the probability of occurrence is used as an ordinate. This figure is useful for describing hardware and people. In the main, failures that occur in the left-hand section of the figure, during early life, are due to problems arising during the manufacturing (gestation) period or to the system's inability to withstand the ambient environment in which it is placed at the beginning of its life. This portion of the distribution includes those items (infants) which may already be in a failed state when they are turned on (born) at $t = 0$. Sometimes this type of failure manifests itself shortly after $t = 0$. This failure mode may be used as a basis for guarantees and warranties for hardware systems. It is taken into consideration for human systems by insurance companies, who do not include newborn children in family health insurance plans until after one or two weeks have passed. Actuarial mortality tables indicate that congenital or hereditary defects may cause death as long as ten years after birth.

The "chance failure" portion of the curve represents a more homogeneous population, with a fairly constant number of failures per unit time and a failure rate that is generally lower than in other portions of the curve. "Infant mortality" of hardware systems can be decreased by aging ("burning in") the individual elements prior to their installation in the system. The final section of the curve, the "wearout phase," is characterized by an increasing

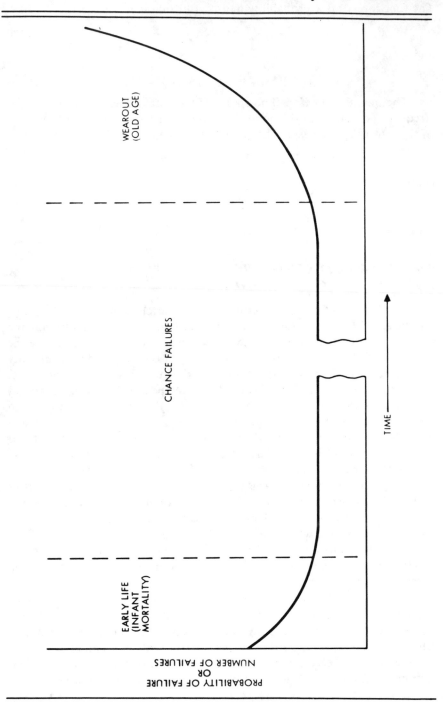

Figure 9-3
General Distribution of Failure Occurrence

WEAROUT
(OLD AGE)

CHANCE FAILURES

EARLY LIFE
(INFANT
MORTALITY)

TIME

PROBABILITY OF FAILURE
OR
NUMBER OF FAILURES

number of failures per unit time, a process that continues until the entire population has failed. Causes of wearout include depletion of material needed for system operation and an increasing inability of the system to use or synthesize necessary materials or energy needed for continued existence. Typical causes of such faults include an accumulation of fatigue, waste materials, or poisons.

The chance failure portion of a system's life cycle is its safest period, since it has the lowest failure rate per unit time. The safety level of a hardware system can be improved, therefore, by using the system after burn-in of its elements and by preventing its use after the onset of wearout.

9.6 COMMONLY USED PROBABILITY FUNCTIONS

The probability density and mass functions that describe most random events of interest in the safety domain (and many in reliability, maintainability, and other domains) are presented in the left-hand columns of Figures 9-4a and 9-4b. The survival and hazard rate functions in these figures are discussed in later sections of this chapter.

9.6.1 The Normal Probability Density Function

Undoubtedly, the best known of the probability density functions is the Gaussian (normal) distribution. Although named after the German mathematician Karl Gauss (1777–1855), it was first developed by Abraham Demoivre in 1733 in connection with studies he made on games of chance.

There are few random phenomena in nature that are precisely described by the normal distribution. Maxwell's law, which specifies the velocity of a molecule for any given direction as a function of temperature, is an example of a random phenomenon that is described by this distribution. Another distribution describes hazards brought about by wearout of hardware or human elements, the right-hand portion of Figure 9-3. The usefulness of the normal distribution is that its many properties make it convenient to manipulate and that it may be used, in one form or another, to approximate other distributions. For example, the normal distribution can be used as an approximation for all of the probability density functions shown in Figure 9-4a and 9-4b, except for the rectangular. In some instances the approximations use the entire distribution, and in other cases, such as for the Weibull when $\gamma = 1$, only a portion of the normal distribution is needed. In the latter instance, it is necessary to make an adjustment such that the area under the curve remains equal to unity. The use of the normal distribution as an approximation of the binomial is discussed in Section 9.6.5.

The convenience with which the normal distribution can be manipulated may make it desirable, when dealing with a random phenomenon that does

Figure 9-4a
Density, Safety Function, and Hazard Rate of the Normal, Exponential, Gamma, and Weibull Distributions

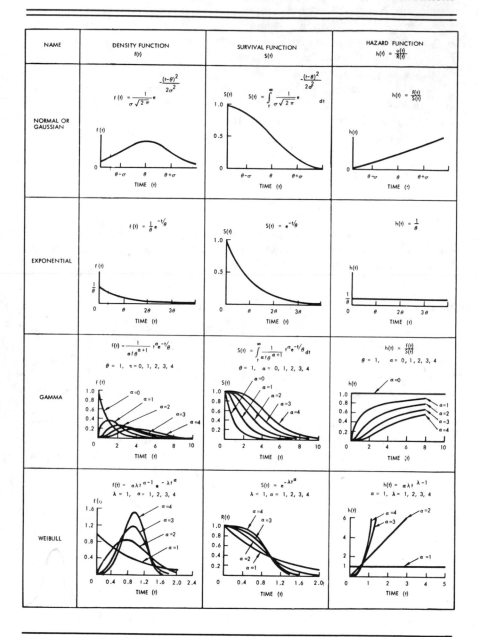

Figure 9-4b

Density, Safety Function, and Hazard Rate of the Rectangular, Binomial, and Poisson Distributions

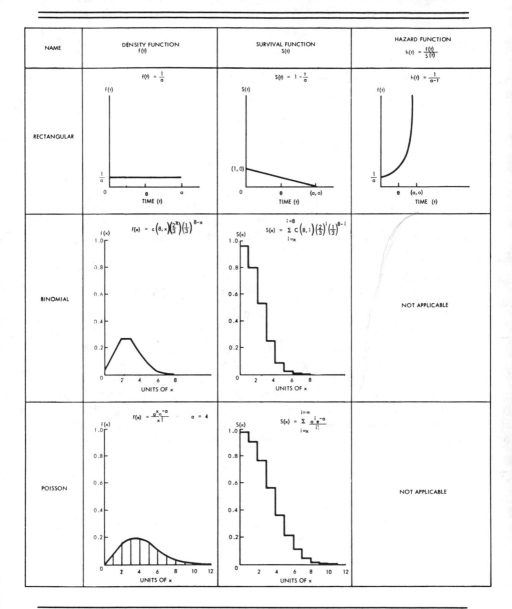

not obey normal probability laws, to make a numerical transformation such that the quantity resulting from the transformation does obey normal probability laws. For example, the natural log of current flowing through an electronic device when a failure is induced may follow normal probability law when voltage does not.

The expression for normal probability density is

$$f(t) = \frac{1}{\sigma\sqrt{2\pi}} \exp\left[-\frac{(t-\Theta)^2}{2\sigma^2}\right]$$
$$-\infty < t < \infty$$

where

$$\sigma > 0$$

(9.11)

where θ and σ are chosen so as to be equal to the first moment and the square root of the second central moment, the mean and standard deviation of this probability law. A discussion of moments is presented in Section 9.7.

9.6.2 The Gamma and Exponential Density Functions

The gamma and exponential probability density functions are also used to describe phenomena, such as hazards and failures, that are randomly distributed with respect to time. The center section of Figure 9-3 obeys exponential probability laws. The exponential distribution may be considered a special case of the gamma density distribution, developed to aid in determining how long an interval of time is required in observing a sequence of events occurring in time (in accordance with a Poisson probability law) at a rate of $1/\Theta$ events per unit time, before the $(\alpha + 1)$ occurrence of the event can be observed. The gamma density function,

$$f(t) = \frac{1}{\alpha!\,\Theta^{(\alpha+1)}} t^\alpha e^{-t/\Theta} \quad,$$

(9.12)

reduces to the exponential when $\alpha = 0$. Consequently, the quantity Θ represents the mean time between occurrences. The exponential distribution is often written

$$f(t) = \lambda \exp[-\lambda t] \quad \text{for } t \geq 0$$
$$= 0 \qquad\qquad \text{otherwise,}$$

(9.13)

where $\lambda = 1/\Theta$ and represents the failure rate of the system.

The gamma and exponential probability laws have recently assumed a greater importance in the safety domain, for these distributions describe numerical-valued random variables such as the time intervals between explosions in mines, the time intervals between successive breakdowns of electronic systems, and the life span of certain electronic components.

9.6.3. The Weibull Density Function

The Weibull probability density function may be expressed as

$$f(t) = \alpha \lambda t^{(\alpha-1)} e^{-\lambda t} \quad \text{for } \alpha, \lambda > 0$$
$$t \geq 0 \tag{9.14}$$

where

α determines the scale of the distribution
λ determines the shape of the distribution

The exponential distribution is a special case of the Weibull obtained by setting $\lambda = 1$. The shape parameter is of interest in the Weibull distribution because it describes the mode of event occurrence as a function of time. When $\lambda = 1$, events occur uniformly over time; a value of $\lambda < 1$ implies that events occur in a decreasing fashion as time increases; and a value of $\lambda > 1$ implies that events increase in number as time increases.

9.6.4 Rectangular Probability Density Function

The rectangular (uniform) probability density function may be expressed as

$$f(t) = \frac{1}{b-a} \quad \text{for } a < t < b$$
$$= 0 \quad \text{otherwise} \tag{9.15}$$

The notion of a uniform probability distribution may be defined as follows. Consider a numerical-valued random phenomenon whose values lie in the finite interval I. That is, $I = $ (real numbers t: $a \leq t \leq b$) for real, finite numbers a and b. The random phenomenon of concern obeys a uniform probability law over I if the value of $P(B)$ of its probability function, for any interval $B \subset I$, satisfies the relation

$$P(B) \quad \frac{\text{length of } B}{\text{length of } I} \quad \text{if } B \subset I$$
$$= 0 \quad \text{if } B \cap I = \Phi \tag{9.16}$$

9.6.5 The Binomial Distribution

The probability functions discussed thus far are continuous and are therefore density functions. The binomial and the Poisson are discrete; consequently, they are mass probability functions. The expression for the binomial may be written

$$f(x;n,p) = \binom{n}{x} p^x q^{n-x} \quad \text{for } x = 0, 1, 2, \ldots, n$$
$$q = 1 - p \qquad (9.17)$$

In the example shown in Figure 9-4b, $n = 8$. The quantity p represents the probability of the undesirable event occurring on a single trial for the case that n trials are attempted.

Equation 9.17 applies to problems that do not involve equally likely descriptions. The value of p is obtained, therefore, by using a frequency definition of probability. Even though it is not feasible to compute a value for p once an estimate is formulated, Equation 9.17 may be used to determine the value for the probability of occurrence of an event S in terms of independent trials of the event A. For example, let S be the probability that three main circuit breakers fail in a group of 100 houses, where a failure in any one house is the event A. The probability of occurrence of the A event may be estimated by statistics gathered by fire underwriters.

Equation 9.17 may also be used to determine the probability that three heads will result (event S) as a result of tossing a coin ten times (event A). This is a special case of Equation 9.17 in which the probability of the event A occurring is constant for each of the n trials and P($\cdot$) is given by

$$P(A_k) = \binom{n}{k} p^k (1 - p)^{n-k} \qquad (9.18)$$

This occurs, for example, when calculating the probability of drawing a white ball from an urn containing m balls, of which n are white, when the ball drawn is replaced before the next drawing. In the safety domain Equation 9.18 is applicable to a small ensemble of systems when each failed system is returned to its original condition subsequent to the occurrence of a hazardous event.

The quantity f of Equation 9.17 is tabulated for various values of p, for $n = 2, 3, \ldots, 49$ in National Bureau of Standards, *Tables of the Binomial Probability Distribution*, and for $n = 50, 51, \ldots, 100$ in Romig, *50–100 Binomial Tables*, (see Additional Reading). Aside from the fact that modern computers tend to make compiled tables obsolete, there is no great need to extend these tables for large values of n. If the binomial probabilities $f(x;n,p)$ are plotted for values of n as a function of x, continuous curves drawn through the histograms approximate the normal density function more accurately as n increases. This quality of the normal distribution is discussed in Section 9.6.1. As a general rule, it is acceptable to use a normal approximation to binomial probability law when $np(1-p) > 3$.

9.6.6 The Poisson Mass Distribution

The Poisson probability law was first published in 1837 by the French mathematician Siméon Denis Poisson (1781–1840) in his book *Recherches sur la probabilité des jugements en matière criminalle et en matière civile*. In a sense, this was the first distribution to be used in the safety domain because it was applied in the nineteenth century to phenomena such as deaths among Prussian soliders resulting from being kicked by horses and to suicides among women and children.

It has since been observed that the Poisson probability law may be used to describe a variety of random phenomena, such as electron emission from heating element filaments and photosensitive substances subjected to light, noise in electronic devices, and spontaneous decomposition of radioactive atomic nuclei. In human affairs, the Poisson distribution is useful for describing demands for service, such as telephone switchboards, automobile tollbooths, stock on hand in department stores, and food supplies available in large cities. The Poisson probability mass function may be written

$$P(x) = \frac{\lambda^x}{x!} e^{-\lambda} \quad \text{for } x = 0, 1, 2, \dots, n$$
$$= 0 \qquad \text{otherwise} \tag{9.19}$$

In application, this distribution is used to determine the number of events where large numbers of opportunities exist for an event to occur, but there is a small probability that any one of the opportunities will result in the occurrence of the specified event.

Specifically, if λ is set equal to μt, where μ is the number of events occurring in a period of time t (or space, length, area, or volume), Equation 9.19 gives the probability that exactly x events will occur in the unit t. Consequently, μ is the mean rate of occurrences per unit t, in the sense that the number of events occurring in an interval $t = 1$ obeys Equation 9.19 with a mean μ. When p is very small and n is large, the terms

$$\binom{n}{x} p^x q^{n-x}$$

can be approximated by

$$\frac{\lambda^x e^{-\lambda}}{x},$$

where $\lambda = np$. In this case, the Poisson distribution may be considered to be a limiting form of the binomial distribution. As a general rule the Poisson expression may be used for the binomial when $p < 0.1$. It was noted in the previous section, however, that the binomial distribution may be approximated by the normal when n large. If p is small enough so that the Poisson distribution may be used to approximate the binomial, but n is large enough so that the normal distribution may be used to approximate the binomial, then the Poisson and normal distributions approximate each other for large values of $\lambda = np$.

The kinds of random phenomena that lend themselves to description by the Poisson probability law are similar to those that are describable by the binomial law. The binomial applies when n independent occurrences are observed. From these one can determine the number of trials in which a certain event occurred and, consequently, the number of trials in which the event did not occur. As indicated in Section 9.6.5, the value of p in the binomial distribution, Equation 9.17, needs to be determined empirically or by deduction. There are certain types of events, however, which cannot be considered to be the results of conducting trials but which must be considered in terms of events per unit time (or space, volume, and so on). For example, suppose one observes the number of accidents occurring at a given intersection for a given interval of time or the number of hostile organisms per unit volume of blood fluid. Given such data, it is not possible to discuss the number of accidents that did not occur in that interval of time at the given intersection, and it makes no sense to consider the number of organisms not present per unit volume of blood fluid. Consequently, the value of λ in Equation 9.19, which is analogous to np of the binomial, cannot be determined theoretically but must be estimated or obtained through observation, as is required for p in the binomial.

Tables of $p(x, \lambda)$ for values of λ between 0.1 and 1.5 may be found in Biometrika tables, and more extensive tables have been calculated by Molina (see Additional Reading).

9.6.7 Other Probability Mass Functions

Other useful probability mass and density functions, such as the Bernoulli, geometric, negative binomial, and the hypergeometric, have characteristics that dictate the circumstances when each is most useful. For example, the Bernoulli random variable can assume one of two values, 0 or 1, that correspond to safe or unsafe or success or failure. The probability mass function for the Bernoulli probability law is

$$
\begin{aligned}
p(x) &= p && \text{if } x = 1 \\
&= 1 - p = q && \text{if } x = 0 \\
&= 0 && \text{otherwise}
\end{aligned}
\tag{9.20}
$$

where $0 \leq p \leq 1$. The outcome of a Bernoulli trial, in which the probability of being safe (success) is p, is an example of a numerical-valued random phenomenon obeying the Bernoulli probability law. As Equation 9.20 shows, x can only take on values of 0 or 1. The binomial probability distribution, Equation 9.17, describes the case in which n independent Bernoulli trials are made and in which the probability of success p is not restricted to exactly 0 or 1.

The geometric probability law for $0 \leq p \leq 1$ is given by the probability mass function

$$p(x) = p(1-p)^{x-1} \quad \text{for } x = 1, 2, \ldots, n$$
$$= 0 \qquad\qquad \text{otherwise} \qquad\qquad (9.21)$$

This distribution is applicable for calculating the number of trials required to obtain the first success in a sequence of n independent Bernoulli trials when the probability of success of each trial is given by p. The negative binomial probability distribution with parameters p and r, where $0 \leq p \leq 1$, $r = 1$, $2, \ldots$, and $p = 1 - q$, is given by the probability mass function

$$p(x) = \binom{r+x-1}{x} p^r q^x = \binom{-r}{x} p^r (-q)^x \text{ for } x = 0, 1, \ldots$$
$$\qquad\qquad\qquad\qquad\qquad\qquad\qquad\qquad (9.22)$$
$$= 0 \qquad\qquad\qquad\qquad\qquad \text{otherwise}$$

This distribution is applicable when obtaining the number of failures encountered before the rth success occurs in a sequence of independent Bernoulli trials where the probability of success of each trial is equal to p. As Equation 9.22 implies, the number of trials required to achieve the rth success is equal to r plus the number of failures encountered prior to achieving the rth success. When $r = 1$, Equation 9.22 reduces to the geometric probability law, Equation 9.21.

As indicated in Section 9.6.5, the binomial distribution is applicable when p is constant, such as occurs when initial conditions are restored after each Bernoulli trial. The hypergeometric distribution, which like the Poisson is a chief example of non-Bernoulli sampling, is applicable when the initial conditions are not restored after each trial. The hypergeometric probability law is given by the probability mass function

$$p(x) = \frac{\binom{Np}{x}\binom{Nq}{n-x}}{\binom{N}{n}} \quad \text{for } x = 0, 1, \ldots, n$$
$$= 0 \qquad\qquad\qquad\qquad \text{otherwise} \qquad\qquad (9.23)$$

where

$$N = \text{any integer } 1, 2, \ldots$$
$$n = \text{an integer in the set } 1, 2, \ldots, N$$
$$p + 1 - q = 0, 1/N, 2/N, \ldots, 1.$$

An example of an application of the hypergeometric distribution is the requirement to calculate the number of defective devices, selected when n safety devices are drawn without replacement, from a population of size N when it is known that there are exactly Np defectives in the group of N devices.

9.6.8 Extreme Value Distribution

The extreme value distribution, a continuous probability distribution, is useful in describing failure modes when it is convenient or appropriate to judge a sample by the largest (or smallest) value observed. This distribution may be applicable to the design of structures, such as dams, buildings built in earthquake regions, tunnels, aircraft wings, and components intended for surgical implantation in humans. It may also be used for analysis of the breaking strength of textiles, meteorological extremes such as rainfalls, floods, and droughts, the span of human life, and breakdown voltage of electronic components.

One limitation in using the extreme value probability law is the need to satisfy the requirement that all observations constitute an independent random sample from the same population. (The risk involved in rejecting a given datum from a set of observations is discussed in the next section.) The extreme value probability density function is given by

$$f(x) = ae^{-a(x-u)-e^{-a(x-u)}}$$

$$a = 1/\beta > 0 \tag{9.24}$$

where u is the highest point (the mode) of the frequency distribution and β is a scale parameter analogous to the standard deviation σ of the normal. (β equals $\sqrt{6}/\pi$ times the standard deviation of the extreme-value distribution.) The general shape of the extreme value density function is shown in Figure 9.5.

Although the parameters u and β specify Equation 9.24, it may be convenient to introduce ξ, a combination of u and β in the form $\xi = u + \beta y$, where known values are assigned to y. In Figure 9-5 the area P, occurring to the left of ξ_p, represents the probability P that a value larger than ξ will not occur. If ξ is very large, the value of P approaches unity. Therefore, ξ_p

Figure 9-5
General Shape of Extreme Value Density Function

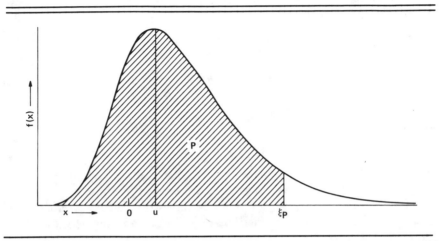

represents the estimate, for various probability levels, such as $P = 0.90, 0.95$, and 0.99, that the observed value of x at the point ξ_p will, on the average, be exceeded 10 times, 5 times, and once, respectively, out of 100 opportunities.

9.6.9 Acceptance or Rejection of Extreme Observations

The question may arise in an experiment requiring probabilistic considerations as to whether certain of the extreme values obtained as part of the data collected are "stragglers" and should be discarded or whether, although extreme, the values are valid and should be included in the calculations. If it may reasonably be assumed that the data collected come from a normal population, then the risk of rejecting an individual datum of extreme value can be derived from tables whose argument permits calculating the probability that the largest (or smallest) item in a sample of size n, gathered from a standard normal population, is less than x. (See Fisher and Tippett; Pearson and Hartley in Additional Reading.) The tables are based upon the following rationale. Let the observed values of x_i in the sample space be arranged in ascending order

$$x_1 \leq x_2 \leq \ldots \leq x_n$$

If $F(x)$ is the distribution function of the parent population, the probability that $x_n \leq x$ is given by

$$[F(x)]^n.$$

This is true because if $x_n \leq x$, the same inequality must hold for all other values in the sample which, however, are supposed to be independent. The probability that the largest item in a sample size N, taken from a standard normal population will be less than x is given by

$$P(x_n \leq x) = \left[\phi(x) \right]^N$$

where

$$\phi(x) = \sqrt{\frac{1}{2\pi}} \int_{-\infty}^{x} e^{-u^2/2} \, du$$

(9.25)

A brief extract of the tables showing values of the upper (and lower) percentage points of the largest value x is presented in Table 9.1

As an example of the way in which the tables are used, consider a circuit breaker intended to interrupt the flow of current at an average value of 50 amperes, with a standard deviation of not more than 0.8 amperes. If sample sizes of 20 circuit breakers are tested routinely, the highest values obtained should not be above $50 + [(0.8)(3.289)] = 52.63$ more than once in a hundred times. If a value greater than 52.63 amperes is observed more frequently, it would be appropriate to determine the cause.

When the character of the parent distribution is not known for certain or it is known to be other than the normal distribution, the criteria of Equation 9.25 is not applicable. The reader is directed to the Additional Reading; see, for example, E. S. Keeping, for a discussion of criteria that may be used in this case.

Table 9-1
Upper Percentage Points of Largest x_N

Sample Size N	Upper Percentage Points	
	5%	1%
5	2.319	2.877
10	2.568	3.089
15	2.705	3.207
20	2.799	3.289
30	2.929	3.402
50	3.082	3.539
100	3.283	3.718
1000	3.884	4.264

9.7 *MOMENTS OF A PROBABILITY FUNCTION*

Given a set of observations about an unsafe condition, the ability to estimate the possibility of a next occurrence depends in part upon the accuracy with which the data can be fitted to a known probability distribution. Knowledge about a distribution implies that the values of the parameters that define the distribution, such as Θ and σ in the normal distribution or n and p in the binomial distribution, are known with reasonable precision. In many instances, however, the values of these parameters are not precisely known and must be estimated. One method of estimation makes use of the relation between the moments of the data collected and the moments of the population from which the data were obtained.

To summarize the characteristics of a probability function $P(\cdot)$ that describes the universe of random phenomena relating to some hazardous event for which a sampling of data is available, consider the notion of the expectation of a continuous function $g(x)$ of a real variable x with respect to the probability function $P(\cdot)$. This expectation $E[g(x)]$ has essentially the same properties as the average of $g(x)$ with respect to a set of numbers. $E(x)$, the expected value of the variable, is known as the "mean of the probability law" or the ensemble average. For a discrete probability law with mass function $p(\cdot)$,

$$E(x) = \Sigma x p(x)$$

over all x such that $p(x) > 0$. (9.26)

For a continuous probability law, with probability density function $f(\cdot)$,

$$E(x) = \int_{-\infty}^{\infty} x f(x) dx \qquad (9.27)$$

The mean of a probability law can be interpreted as follows. Suppose x_1, $x_2, \ldots, x_n, \ldots$, are independent data points of the random phenomena under consideration. Successive arithmetic means are formed from the succession

$$m_1 = x_1$$

$$m_2 = \frac{x_1 + x_2}{2},$$

$$m_3 = \frac{x_1 + x_2 + x_3}{3}, \ldots,$$

$$m_n = \frac{x_1 + x_2 + \ldots + x_n}{n}, \ldots,$$

and the limiting value of this succession is the mean of the probability law.

The expectation of the function $g(x) = x^2$, written $E(x^2)$, the second moment of the probability law, is referred to as the mean square of the probability law. (It should be noted that $E(x^2)$ is not the same as the square of the mean of the probability law, $E^2(x)$. For a discrete law, with a mass function $p(\cdot)$,

$$E(x^2) = \Sigma x^2 p(x)$$

over all x such that $p(x) > 0$ \hfill (9.28)

For the continuous case, with probability density function $f(\cdot)$,

$$E(x^2) = \int_{-\infty}^{\infty} x^2 f(x)dx$$ \hfill (9.29)

Extending this to the general case, $E(x^n)$ is the expectation of $g(x) = x^n$ and is the nth moment of the probability law, where, as mentioned earlier, the first and second moments are the mean and mean square of the probability law. The first and second moments of the uniform probability density function, Equation 9.15, are given by

$$E(t) = \int_{-\infty}^{\infty} tf(t)dt$$

$$= \frac{1}{b-a} \int_{a}^{b} tdt = \frac{b^2 - a^2}{2(b-a)} = \frac{b+a}{2}$$

and

$$E(t^2) = \int_{-\infty}^{\infty} t^2 f(t)dt$$

$$= \frac{1}{b-a} \int_{a}^{b} t^2 dt = \frac{b^2 + ba + a^2}{3}$$

The first and second moments of the Bernoulli mass function, Equation 9.20, are given by

$$E(x) = 0 \times q + 1 \times p = p$$

$$E(x^2) = 0^2 \times q + 1^2 \times p = p$$

For any real number c and integer $n = 1, 2, \ldots$, $E[(x-c)^n]$ is called the nth moment of the probability law about the point c. Of particular interest is the case when $c = E(x)$. $E[x - E(x)]^2$ is of particular importance and is known as the variance of the probability law. The variance and the mean are the two most widely used moments. Using m and σ^2 for the mean and variance, respectively, they are written

$$m = E(x)$$ \hfill (9.30)

$$\sigma^2 = E[(x-m)^2] = E(x^2) - E^2(x)$$ \hfill (9.31)

where σ is referred to as the standard deviation of the probability law.

As an example, consider the variances of the uniform probability density function and the Bernoulli probability mass function. These are given, respectively, by

$$\sigma^2 = E(x^2) - E^2(x) = \frac{b^2 + ba + a}{3} - \frac{(b+a)}{2}$$

$$= \frac{(b-a)^2}{12}$$

$$\sigma^2 = E(x^2) - m^2 = p - p^2 = p(1-p) = pq$$

Values for m and σ^2 for some of the various probability mass and density functions discussed thus far are presented in Tables 9-2 and 9-3, respectively.

9.8 CHEBYCHEV'S INEQUALITY

Suppose data are gathered about a set of hazardous occurrences so that a mean and variance can be obtained, but not the functional form of the probability law describing the general circumstance. One procedure for making predictions about the general case is to presume a probability law that describes the universe. The reasonableness of such a presumption depends upon the amount of data collected and upon the closeness with which the data fit the probability law presumed. Another procedure is to apply a nonparametric assessment to determine the intervals in which, with a

Table 9-2
Mean and Variance for Discrete Probability Functions

Law	Parameters	Mean	Variance
Bernoulli	$0 \le p \le 1$	p	pq
Binomial	$n = 1, 2, \dots$ $0 \le p \le 1$	np	npq
Poisson	$\lambda > 0$	λ	λ
Geometric	$0 < p \le 1$	$1/p$	q/p^2
Negative binomial	$r > 0$	$\dfrac{rq}{p} = rp$	$\dfrac{rq}{p^2} = rPQ$
	$0 < p \le 1$	if $P = \dfrac{q}{p}$	if $Q = \dfrac{1}{p}$
Hypergeometric	$N = 1, 2, \dots$ $n = 1, 2, \dots, N$ $p = 0, \dfrac{1}{N}, \dfrac{2}{N}, \dots, 1$	np	$npq \left(\dfrac{N-n}{N-1} \right)$

Table 9-3
Mean and Variance for Continuous Probability Functions

Law	Parameters	Mean	Variance
Uniform over interval a to b	$-\infty < a < b < \infty$	$\dfrac{a+b}{2}$	$\dfrac{(b-a)^2}{12}$
Normal	$-\infty < m < \infty$ $\sigma > 0$	m	σ^2
Exponential	$\lambda > 0$	$\dfrac{1}{\lambda}$	$\dfrac{1}{\lambda^2}$
Gamma	$r > 0$	$\dfrac{r}{\lambda}$	$\dfrac{r}{\lambda^2}$

high probability, an observed value of the numerical-valued random phenomena under consideration may be expected to lie. One such nonparametric criterion is based on Chebychev's inequality, named for the Russian probabilist P. L. Chebychev (1821–1894). To apply this inequality, consider the quantity $Q(h)$, for any $h > 0$, defined as the probability assigned to the interval $[X: m - h\sigma < x \leq m + h\sigma]$ by a probability law. In terms of the distribution function $F(\cdot)$ or the probability density function $f(\cdot)$,

$$Q(h) = F(m + h\sigma) - F(m - h\sigma)$$

$$= \int_{m-h\sigma}^{m-h\sigma} f(x)dx \tag{9.32}$$

$Q(h)$ can be calculated from Equation 9.32 for the various density functions discussed thus far. For example, $Q(h)$ for the exponential law with $m = 1$ is given by

$$Q(h) = e^{-1}(e^h - e^{-h}) \qquad \text{for } h \leq 1 \tag{9.33a}$$

and by

$$Q(h) = 1 - e^{-(1+h)} \qquad \text{for } h \geq 1 \tag{9.33b}$$

For the uniform distribution over the interval a to b for $h < \sqrt{3}$

$$Q(h) = \frac{1}{b-a} \int_{[(b+a)/2] - h(b-a)/\sqrt{12}]}^{[(b+a)/2] + h(b-a)/\sqrt{12}]} dx = \frac{h}{\sqrt{3}}$$

It can be shown that $Q(h)$ has a lower bound known as Chebychev's inequality, which is independent of any explicit probability law. For the continuous case,

$$\int_{m-h\sigma}^{m+h\sigma} f(x)dx \geq 1 - \frac{1}{h^2} \qquad (9.34)$$

Restating the inequality in terms of an observed value x of a numerical-valued random phenomenon, $Q(h)$ is equal to the probability that an observed value of a numerical-valued random phenomenon, with a density function $f(\cdot)$, will lie in an interval centered at m whose length is $2h$ standard deviations. That is,

$$Q(h) = P(|x-m| \leq h\sigma)$$

For $h > 0$ Chebychev's inequality may be written

$$P(|x-m| \leq h\sigma) \geq 1 - \frac{1}{h^2}$$

$$P(|x-m| > h\sigma) \leq \frac{1}{h^2} \qquad (9.35)$$

Applying Equation 9.34, it can be seen that when $h=2$, the probability is at least 0.75 that an observed value x will fall within 2 standard deviations of the mean, and when $h=5$ the probability is at least 0.96 that x will fall within 5 standard deviations of the mean.

9.9 SURVIVAL PROBABILITY FUNCTION

The distribution function $F(\cdot)$ defined in Section 9.2 may now be restated as follows. There is a probability distribution function $F(x)$, associated with the outcome of an experiment or with data obtained from a set of observations about some hazardous circumstance, such that observations made of the random variable x' can be ordered in terms of likelihood of occurrence. That is, the probability of the event $x' \leq x$ is expressed by $F(x)$, Equation 9.1. Applying both Equations 9.1 and 9.2, the probability that x' falls between two specified observations x_1 and x_2 may be determined from

$$F(x_2) - F(x_1) = P(x_1 < x' \leq x_2) \qquad (9.36)$$

It may be of interest to be able to assess the limits of the distribution which define the safe event S. If x_1 and x_2 define the limits in $F(x)$ that bound S, and $\Theta_1, \Theta_2, \ldots$ are the parameters which describe the distribution $F(x)$, S is given by

$$S = P(x_1 < x' \le x_2) = F(x_2 : \Theta_1, \Theta_2. \ldots) - F(x_1 : \Theta_1, \Theta_2. \ldots)$$

$$= S (\Theta_1, \Theta_2, \ldots). \tag{9.37}$$

This is referred to as the *survival* or *reliability* function. Survival functions defined by Equation 9.37 are applicable, for example, to strength of medications and vaccines, the concentration of certain chemicals in blood, and the amount of energy required to activate safety devices such as fuses and pressure valves.

The survival function expressed in Equation 9.37 is also applicable to certain distributions such as the normal and exponential. In the case of the exponential distribution, the survival function may be defined in terms of the probability that the system will operate for a period of time equal to or greater than T. That is,

$$S(\theta) = P(\mathrm{T} > T) = \int_T^{\infty} \frac{1}{\Theta} e^{-t \cdot \theta} \, dt = e^{-T \cdot \theta} \tag{9.38}$$

Operation for an interval of time that is at least equal to T is useful for survival considerations of transportation systems such as cars, trains, and aircraft and of humans. For these systems, it may also be desirable to describe the survival function in terms of the probability that no Class I hazard will occur in some interval of time T. This may be written:

$$S(t) = P(0,t) = 1 - P(x \ge 1, t) \tag{9.39}$$

where, if x is the number of Class I hazards occurring in time t, $P(x \ge 1, t)$ is the probability of one or more failures occurring in time t.

A probability mass distribution may also be used as a basis for determining a survival function. For example, let Θ be the probability of one system element causing a hazard to the system, where hazardous occurrences of the system elements follow a binomial distribution. Then the survival bfunction may be defined as the instance when at least r components out of a total of n function satisfactorily. That is,

$$S(\theta) = P(r \le \xi \le n)$$

$$= \binom{n}{n} \theta^n + \binom{n}{n-1} \theta^{n-1}(1-\theta) + \ldots + \binom{n}{r} \theta(1-\theta)^{n-r} \tag{9.40}$$

The survival functions for several of the most commonly used probability density and mass functions are presented in the center column of Figures 9-4a and 9-4b. The survival function for the general distribution presented in Figure 9-3 is presented in Figure 9-6. The quality common to all survival functions is that the probability of survival is presumed to be unity at $T = 0$. At anytime subsequent to $T = 0$ the form of the survival function depends upon the form of the probability density or mass function.

Figure 9-6
Survival Function for General Failure Distribution

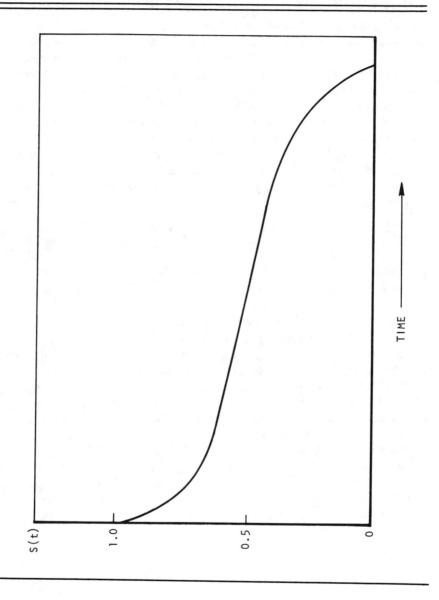

9.10 HAZARD PROBABILITY FUNCTION

The *hazard function* $h(t)$ of the probability distribution that describes time to hazardous occurrence may be defined as the conditional probability density function of time to hazardous occurrence, given that no hazard has occurred to the system prior to time t. If, as discussed in Section 9.4., $f(t)$ is a probability density function, then $f(t)dt$ represents the proportion of hazardous occurrences that occurs to the system and its users during the interval $(t, t+dt)$, presuming that system operation was initiated at $t=0$. Hazards occurring during some interval of time prior to $(t, t+dt)$ are described in a similar fashion. Under this condition, $h(t)dt$ represents the proportion of hazardous occurrences that occur during the interval $(t, t+dt)$, but which have not occurred prior to t. The relationship between $f(t)$ and $F(t)$ and $h(t)$ can be shown to be

$$h(t) = \frac{f(t)}{1 - F(t)} = \frac{f(t)}{S(t)} \tag{9.41}$$

where $F(t)$ and $f(t)$ are defined in Sections 9.2 and 9.3, respectively.

The distinction between the hazard rate $f(t)$ and the instantaneous hazard rate $h(t)$ may be illustrated by the following analogy. If an automobile trip of 100 miles is completed in 2 hours, the average speed is 50 mph, with actual speeds varying from 0 mph to some value greater than 50 mph. The average value, 50 mph, is analogous to the hazard rate obtained from the probability density function, while the speed at any instant of time is analogous to the instantaneous hazard rate. In relation to humans, the hazard rate of the general distribution, shown in Figure 9-3, is analogous to the death rate while the instantaneous hazard rate is the force of mortality, as it is referred to in actuarial theory.

The instantaneous hazard distribution function is shown in the right-hand column of Figure 9-4a and 9-4b for the set of probability density distribution functions presented in these figures. The instantaneous hazard rate is not applicable to the probability mass distribution functions of Figure 9-4b.

Circumstances may arise in safety when the distribution function describing the hazard is not known but when the data collected provide knowledge about the instantaneous hazard rate. Since by Equation 9.2, $F(\infty)=1$,

$$\int_0^t h(x)\,dx \to \infty$$

as $t \to \infty$. $F(0)$, however, is permitted to have any value between 0 and 1. If $h(t)$ is known and $F(0)$ is assumed to be equal to 0, a reasonable assumption, then Equation 9.41 can be solved for $F(t)$ and $f(t)$, yielding

$$F(t) = 1-\exp\left[-\int_0^t h(x)dx\right] \qquad (9.42a)$$

$$f(t) = h(t)\exp\left[-\int_0^t h(x)dx\right] \qquad (9.42b)$$

respectively. Consider, as an example, the case where

$$h(t) = \lambda \quad \text{for } \lambda > 0 \qquad (9.43)$$

Applying Equation 9.42b, it can be seen that

$$f(t) = \lambda e^{-\lambda t}, \qquad (9.44)$$

which is the exponential probability density function, Equation 9.13, where λ represents the failure rate and is the reciprocal of the mean-time-to-failure rate. The converse implies that a system has a constant instantaneous hazard rate if its rate of hazardous occurrences is describable by an exponential probability function or by a gamma or Weibull probability density function with α equal to 0 or 1, respectively.

9.11 DISTRIBUTIONS APPLIED TO SMALL SAMPLE SIZES

The general distribution of Figure 9-3 and, by implication, those of Figures 9-4a, 9-4b, and 9-5 presume a large sample size of identical or very similar elements or systems. However, it is not uncommon for just one or a very few systems to be available for testing or to be produced for actual use. Long life is achieved for such systems by making repairs or modifications when problems arise. The case in which there are very few systems for testing can be treated as if there were a large number of samples if ergodicity exists.

> *Ergodicity* is that property of a system which makes it tend, probabilistically, to a limiting form that is independent of the initial position from which it started. That is, time averages for a single system are equivalent to ensemble averages.

To illustrate application of the ergodic hypothesis to a system, consider an idealized billiard ball of mass point and no friction. If placed in motion, this idealized ball will maintain its speed forever and, except for some initial conditions, will in time approach arbitrarily any given point on the table. Consider an ensemble of trajectories, starting from all possible positions and directions in which the probability of starting in any given direction and in any given angular range is proportional to the area of that region multiplied by the angular range. The ergodic hypothesis states that the ensemble average, defined by the position and direction of an ensemble of such trajectories taken at any given instant in time, is equal to the time average obtained for any one trajectory, measured over a period of time that is long enough to provide a sample size equivalent to the one obtained from the ensemble.

Applied to safety, the ability to utilize the ergodic hypothesis implies that the set of unsafe events occurring during operations of a large ensemble of systems, such as a group of automobiles or aircraft, will be the same during a small interval of time as would occur during the equivalent lifetime of one or just a few systems. If applicable, this hypothesis permits test programs, intended to verify hypotheses for an ensemble of systems, to be carried out with the small number of samples, provided

1. The amount of test time applied to the small sample is equivalent to the amount of time that would otherwise be obtained by testing a large number of systems.
2. Modifications or repairs made to the few systems undergoing tests return each system to a condition identical to the state it was in at $t = 0$.

9.12 DISTRIBUTION OF HUMAN PERFORMANCE

No special distinction was made in the discussion of the various probability distributions between their application to hardware and to human systems. It was noted, however, in the previous chapter that there are additional difficulties in predicting occurrence of human system failure as compared to hardware system failure. Although failures, human or hardware, are not necessarily synonymous with unsafe conditions, an increase in the number of failures per interval of time generally increases the probability of occurrence of an unsafe event. Askren and Regulinski have suggested as a model for the reliability of human performance for continuous tasks the function

$$R(t) = \exp\left[-\int_0^t e(t)dt\right]$$

(9.46)

where

$R(t) =$ the reliability of human performance for any point in time
$e(t) =$ the error rate for any specific tasks

Values for $e(t)$ are obtainable through techniques such as THERP, discussed in Section 8.11.

Additional Reading

Askren, W. E., and T. L. Regulinski. Quantifying human performance for reliability analysis of systems. *Human Factors* vol. 2, no. 4, August 1969, pp. 393–396.

Cramer, H. *Mathematical Methods of Statistics*. Princeton, N.J.: Princeton University Press, 1946.

Fisher, R. A., and L. H. C. Tippett. *Limiting Forms of the Frequency Distribution of the Largest or Smallest Member of a Sample*. Proceedings of the Cambridge Philosophical Society, 24 (1928), pp. 180–190.

Gumbel, E. J. *Statistics of Extremes*. New York: Columbia University Press, 1958.

Hald, A. *Statistical Theory with Engineering Applications*. New York: Wiley, 1952.

Keeping, E. S. *Introduction to Statistical Inference*. New York: Van Nostrand, 1962.

Kozelka, R. *Elements of Statistical Inference*. Addison-Wesley Series in Statistics. Reading, Mass.: Addison-Wesley, 1961.

Molina, E. C. *Poisson's Exponential Binomial Limit*. New York: Van Nostrand, 1947.

National Bureau of Standards. *Tables of the Binomial Probability Distribution*. G.P.O., 1949.

National Bureau of Standards. *Tables of Normal Probability Functions*. G.P.O., 1953.

Parzen, E. *Modern Probability Theory and the Applications*. New York: Wiley, 1960.

Pearson, E. S., and H. O. Hartley. *Biometrika Tables for Statisticians*, Table 24. Cambridge, Eng: Cambridge University Press, 1954.

Romig, H. G. 50–100 *Binomial Tables*. New York: Wiley.

Von Alven, W. H. *Reliability Engineering*. Englewood Cliffs, N.J.: Prentice-Hall, 1964.

Whitney, D. R. *Elements of Mathematical Statistics*. New York: Holt, 1959.

APPENDIXES

SYSTEM SAFETY CHECKLISTS

The objective of a system safety program is to aid in achieving one or more predetermined goals, an objective accomplished by conducting a series of tasks and by implementing a set of controls on the nature and the timing of these tasks to assure proper, expeditious completion. The tasks carried out in this way interface and are integrated with the efforts of various organizations, such as design, purchasing, engineering, reliability, quality assurance, and product support, and are carried out during the entire life span of a program, from preliminary concept analysis to operations. Areas of concern to system safety include the following:

1. Design, material, and workmanship
2. Systems and equipment interfaces
3. Scheduled and unscheduled maintenance
4. Human errors of omission or commission
5. Occurrence of credible accidents
6. Natural phenomena

System safety checklists can be an aid in carrying out the tasks required to achieve the safety goals. Checklists are presented in this appendix for management of the following organizations:

1. Program control
2. Product integrity
3. System safety
4. Engineering
5. Testing
6. Quality assurance
7. Sales
8. Product support and field service

All the organizations listed are not necessarily applicable to a given project, and in some instances additional organizations may have to be added. Each list is general in nature and is intended to cover the time span from concept formulation through operations or its equivalent, product use. To ensure completeness some overlap was permitted between the organizations listed and lower levels.

In using the checklists, you may append additional lines to each existing list as needed and mark "not applicable" on those items that do not apply to a particular project.

PROGRAM CONTROL	Yes	N/A
Have the contractual requirements for safety been identified?	____	____
Have the potential risks associated with the contract or product been identified and evaluated?	____	____
Is the safety effort funded in a manner consistent with contractual requirements or product use and any risks identified?	____	____
Has a responsible system safety engineer been appointed for the program?	____	____
If required, has a plan been prepared to deal with unconventional hazards?	____	____
Are there provisions to ensure that effective action will be taken on the findings of system safety analysis?	____	____
If required, has the system safety effort been integrated with Occupational Safety and Health Administration (OSHA) requirements?	____	____
Have all applicable federal, state, and local restrictions and regulations been identified?	____	____

PRODUCT INTEGRITY

	Yes	N/A
Have all potential users of the product been identified?	___	___
Is the philosophy of product design based on requiring safety in use or providing safety in design?	___	___
Are user instructions available?	___	___
Has the record of incidents and liability claims been reviewed for similar products?	___	___
Have inherent, uncontrollable hazards been identified for the user?	___	___
Is product safety information disseminated to all levels of management, supervision, and job performers?	___	___
Are there provisions for customer notification when hazards are identified?	___	___
Is any company-provided safety training adequately described? Have arrangements been made for its presentation?	___	___
Have performance requirements of components and raw materials been provided to suppliers?	___	___
Have arrangements been provided for feedback from product users?	___	___
Have procedures been established for reporting and investigating accidents and mishaps that occur to users?	___	___
Is a centrally controlled accident, mishap, and corrective action file in existence and up to date?	___	___
Are company limits of warranty and guarantee clearly stated?	___	___
Are the responsibilities of suppliers and distributors understood?	___	___

SYSTEM SAFETY	Yes	N/A
Has a safety program been developed for hazard elimination and control?	___	___
Are all safety milestones and their relationships with major program milestones shown?	___	___
Are industry codes and standards adequate for the user's protection?	___	___
Has identification been made of personnel and property hazards that can occur during	___	___
Useful life	___	___
Maintenance and overhaul	___	___
Installation	___	___
Transportation	___	___
The period after useful life	___	___
Product disposal	___	___
Can hazards occur as a result of product misuse?	___	___
Can hazards occur as a result of parts or components manufactured by others?	___	___
Does the safety analysis require inclusion of support equipment, fire, water, emergency systems, emergency electrical systems, etc.?	___	___
Are emergency procedures well defined?	___	___
Is a file provided for all pertinent data affecting the safety program?	___	___
Are provisions included as needed for operations, maintenance, crew training, and user certification?	___	___
Have procedures been established for providing new information to customers? For callbacks?	___	___

ENGINEERING   Yes   N/A

Have codes and standards applicable to the product
been identified?   ____   ____

Are the design control documents containing safety
criteria available?   ____   ____

Does the engineering design philosophy include eliminating,
controlling, or warning of hazards?   ____   ____

Has a mechanism for handling product accident, incident,
and complaint information been defined?   ____   ____

Has a hazard analysis been incorporated into the design of
the product?   ____   ____

Has data flow between system safety and engineering
been identified and described?   ____   ____

Is it clear that hazard avoidance includes hazards to the
system as well as to people?   ____   ____

Have safety criteria also been identified for support
equipment where applicable?   ____   ____

Are there provisions to ensure final action on all serious
hazards prior to product release?   ____   ____

QUALITY ASSURANCE   Yes   N/A

Are safety requirements for transportation and packaging
standards being followed?   ____   ____

Are adequate records maintained of inspection and test?   ____   ____

Has a length of record retention been determined and
implemented?   ____   ____

Have the kinds of records to be gathered from testing,
engineering, and manufacturing been determined?   ____   ____

Are quality control procedures clearly defined and
determined to be adequate?   ____   ____

Are procedures for handling rejects and/or unsatisfactory
products clearly defined and practiced?   ____   ____

Are critical parts permanently identified?   ____   ____

Are protective devices in place before shipment?   ____   ____

Are all product warnings, guarantees, and warranties
in place before shipment?   ____   ____

TESTING

	Yes	N/A
Does the test plan include testing the safety devices and safety design features incorporated in the product and its packaging?	____	____
Is the basic data flow between system safety and testing outlined and understood?	____	____
Is a method of evaluating procedural compliance with safety regulations included?	____	____
Does system safety approve and regulate safety-critical operations?	____	____
Does the testing program provide verification for all advertising and packaging claims?	____	____
Does the testing program provide data needed by regulatory agencies for sale of the product?	____	____
Are data from the field being used in test planning?	____	____

SALES	Yes	N/A
Has the wording of promotional, instructional, and technical data been reviewed to ensure that the product and its use are correctly shown and implied?	____	____
Is product labeling reviewed to ensure that technical figures are correct?	____	____
Is promotional and instructional material presented so as to be clearly understandable by users?	____	____
Has translation of promotional and instructional material into additional languages been considered?	____	____
Are verbal presentations by sales personnel and advertising agencies periodically reviewed to ensure that the product and its use are correctly implied?	____	____
Have all governmental approvals required before sales been obtained?	____	____

PRODUCT SUPPORT AND FIELD SERVICE	Yes	N/A
Are any product service limitations clearly defined and understandable?	____	____
Are the procedures in data flow between system safety and field service documented?	____	____
Does the data flow ensure continuous system safety monitoring of field operations and customer experience?	____	____
Has a review of requirements for operating procedures been made by safety personnel?	____	____
Has system safety approval been obtained for safety-critical operations?	____	____
Are maintenance and training procedures reviewed and updated by safety?	____	____
Are necessary provisions for operations and maintenance training available?	____	____
Have training procedures or user instruction concerning use of the product been established?	____	____
Is training being maintained by product and field personnel to update their knowledge of the system?	____	____
Are all personnel familiar with emergency procedures?	____	____
Are records maintained of communications with product users?	____	____
Are investigations made of product complaints? Are these discussed with safety personnel where appropriate?	____	____
Are procedures for reporting accidents and mishaps understood by all personnel?	____	____

EXPANSION OF CHECKLIST

Appendix A is a checklist suitable for use at high organizational levels. Additional detail is required for planning and executing an actual system safety program. Some illustrations of such an expansion are presented in this appendix for some of the elements in Appendix A.

The expansion was carried out by specifying tasks for several of the life cycle phases shown in Figure 5-2. In addition, a scheme for accountability is suggested. The column labeled percent complete in this appendix can be used in an attribute fashion, with a yes or no indication, or in a variables fashion, showing the portion of the task complete versus the amount scheduled for completion.

PROGRAM CONTROL

1.1 Concept Formulation Phase	Percent Complete	Responsible Party
1.1.1 Allocate necessary funding for system safety participation in the concept formulation phase.		
1.1.2 Prepare a preliminary draft of the System Safety Program Plan.		
1.1.3 Make provisions for system safety participation in conceptual studies to be conducted during this phase.		
1.1.4 Ensure that designs generated in the concept formulation phase are reviewed for system safety requirements.		

	Percent Complete	Responsible Party

1.1.5 Establish a system safety organization for carrying out tasks required during concept formulation.

1.1.6 Review the material prepared in response to the request for proposal to ensure that all requirements established by the proposal are included in the response and are in the proper format.

1.2 Contract Definition Phase

1.2.1 Update the preliminary system safety plan at the completion of Phases A, B, and C.

1.2.2 Assess the extent to which system safety tasks and personnel requirements have been established.

1.2.3 Ensure that the functions, responsibilities, and authorities of system safety have been clearly defined and are implemented by program directives.

1.2.4 Establish a program management group to review the progress of system safety activities on a periodic basis.

1.3 Development and Production Phase

1.3.1 Establish system safety functions in manufacturing, testing, and quality assurance as an integral part of these organizations or by assignment to a central system safety organization.

1.3.2 Provide for system safety monitoring of any development and qualification testing.

	Percent Complete	Responsible Party

1.3.3   Identify safety-critical aspects of test activities and operations which can cause hazards during the design and manufacturing phases.

1.3.4   Establish program requirements for accident/incident investigation and reporting.

1.3.5   Ensure that subcontractor activities are included in the overall system safety requirements.

1.3.6   Establish requirements for training programs necessary to ensure achievement of proper safety levels in the operations phase.

1.3.7   Prepare system safety requirements for handling, packaging, and transportation of components that are potentially hazardous.

1.4 Operations Phase

1.4.1   Prepare system safety program plans for operators of the system and, if necessary, for users.

1.4.2   Issue instructions for system safety requirements to be implemented during the operations phase.

1.4.3.   Establish system safety personnel requirements for carrying out the safety tasks during the operations phase.

1.4.4   Ensure that all safety-critical procedures and operations that take place during the operations phase are identified to operators and users.

	Percent Complete	Responsible Party

1.4.5 Establish requirements for system safety participation in the implementation of recommendations resulting from accident investigations.

1.4.6 Conduct periodic system safety surveys of the various operational sites and operational activities and implement the recommendations resulting from such surveys.

SYSTEM SAFETY

3.1 Concept Formulation Phase

3.1.1 Review the request for proposal and work statements to determine what safety requirements are established explicitly and implicitly for the system.

3.1.2 Review all documents that impinge upon safety and that are made a part of the contract by reference in the request for proposal.

3.1.3 Review internal system safety requirements and government requirements established for systems similar to the one under consideration to determine the extent to which they may be applicable.

3.1.4 Review corporate and divisional safety policies and handbooks to determine whether requirements, in addition to those in the request for proposal, need to be included in the safety program.

	Percent Complete	Responsible Party

3.1.5   Integrate all safety requirements determined to be applicable, and prepare a response to the applicable portions of the request for proposal for use by the program manager.

3.1.6   Initiate the preparation of a preliminary system safety program plan and undertake the necessary steps for its approval by the program manager and the customer.

3.1.7   Submit to the program manager budgetary requirements for carrying out the system safety effort during concept formulation and estimates for the subsequent program phases.

3.1.8   Prepare a set of system safety position descriptions for system safety responsibilities required of other organizations, such as engineering, quality assurance, and operations.

3.1.9   Establish a sequence of system safety tasks suitable for achieving the system safety requirements established by the contract.

3.1.10   Ensure that all the functions and organizations with which systems safety needs to interface during various phases of the system life cycle are briefed on their responsibilities so that the system safety requirements can be achieved. Identify any requirements established for system safety by these organizations so that their objectives can be achieved.

	Percent Complete	Responsible Party

3.1.11 Establish points of contact between system safety and the customer, associate contractors, subcontractors, government agencies, representatives of operators and users, and any other appropriate functions.

3.1.12 Revise system safety policies, procedures, and implementation instructions so as to provide support for subsequent phases of the contract.

3.1.13 Establish conformance between system safety documentation prepared expressly for the system under consideration and policies and guidelines established at the corporate level.

3.1.14 Ensure that system safety inputs have been provided and incorporated into all program plans under preparation with which system safety has an interface, such as reliability and quality assurance.

3.1.15 Determine the requirements and materials necessary for training system safety personnel. Coordinate this activity with organizations whose responsibilities include training.

3.1.16 Outline a set of system safety tasks and establish milestones for their completion for subsequent phases of the program.

3.1.17 Review system safety analyses prepared during this phase of the program to ensure mutual support and agreement between these and other analyses with which they interface.

	Percent Complete	Responsible Party

3.1.18 Review available personnel resources in system safety and other organizations for future selection of personnel to carry out system safety functions in other organizations, such as engineering, testing, and operations.

3.1.19 Review available resources to determine whether required safety tests can be performed internally or whether subcontracts will be required.

3.1.20 Establish methods and procedures needed for ensuring that system safety criteria, standards, and requirements are implemented by the design process.

3.1.21 Participate in trade studies and cost effectiveness studies carried out during this phase of the system life cycle.

3.1.22 Provide system safety requirements for system studies needed from other organizations.

3.1.23 Provide technical safety considerations, conclusions, and recommendations required in technical material needed by the customer.

3.1.24 Assure that system safety inputs to proposals for carrying out subsequent phases of the system life cycle are provided to the customer in proper format and in sufficient detail.

NOTE: Subsequent material in this appendix does not make use of the numbering system established by Tables 4-1 and 4-2 in Chapter 4. However, the headings of the various sections permit the reader to relate the information in this appendix to any matrix that may be developed for describing a system.

	Percent Complete	Responsible Party

Contract Definition Phase

1. Designate an individual to head the system safety effort

2. Review the contract to determine the system safety requirements established therein, explicitly and implicitly.

3. Reach agreement with the program manager on the system safety organization, personnel requirements, budget, and schedules to be fulfilled.

4. Update the system safety program plan to meet the requirements of the contract.

5. Submit system safety task descriptions for subsequent phases of the system life cycle, along with a schedule for their completion, to the program manager for approval.

6. Review the system safety tasks and the schedule for their completion to be conducted by other organizations, such as engineering and operations, to ensure that safety contractual requirements will be met.

7. Initiate preparation of a system safety program plan, hazard analysis studies, determination of the requirements and constraints to be placed in high-level specifications, identification of the data required by safety to carry out its tasks, and preparation of material to be included in subcontracts to be let shortly.

8. Review the operational facilities and equipment and procedures necessary to support identification of potential hazardous conditions that may interface with the system.

	Percent Complete	Responsible Party

9. Participate in system trade-off studies to ensure that the required level of safety is achieved in an optimum manner.

10. Transform the safety requirements and tasks established for the system into requirements and tasks for subsystems and, as appropriate, for lower equipment levels.

11. Perform preliminary operating safety analyses necessary to determine the safety requirements for personnel, procedures, and equipments for use during subsequent system phases.

12. Participate in design reviews of the system, subsystems, equipments, and facilities.

13. Perform special safety reviews needed on systems, subsystems, equipments, and facility designs specified by the contract.

14. Update program plans to reflect changes made to the system safety requirements in the program.

15. Review specifications that have been identified as requiring inputs from system safety.

16. Review associate and subcontractor proposals and specifications for their safety contents.

17. Provide needed inputs to studies, reports, and briefings presented to the program manager and customer.

18. Establish a technical reference file and data bank to support the system safety requirements.

	Percent Complete	Responsible Party

19. Participate in preliminary subcontractor and supplier surveys to ensure their ability to collaborate in achieving the system safety requirements.

20. Check the thoroughness with which system safety standards are used by other functional areas such as engineering, human factors, and reliability.

21. Identify those safety tests which need to be performed. Collaborate with other functional areas performing tests to prevent duplication of effort.

22. Prepare and issue directives to ensure that system safety inputs are provided to laboratory and operation procedures, test plans and procedures, manufacturing production orders, manufacturing handling procedures, material procurement specifications, material purchase orders, material handling, loading and transportation procedures, service manuals, maintenance manuals, and spares provisioning.

23. Submit an estimate to the program manager for the amount of computer time required for subsequent phases of the system life cycle.

24. Revise system safety requirements, standards, and checklists as information is derived from trade-off studies, design development, specification preparation, and evaluations.

25. Review the requirements for data submission with the program manager and the customer to ensure adequate coverage in system safety.

	Percent Complete	Responsible Party

26. Provide recommendations for appropriate effort required in the system safety domain to optimize the response to any incentive provisions contained in the contract.

27. Evaluate the system safety effort to ensure that effective business systems and management techniques and methods are being used, and that they are not being duplicated by other functional areas in the project.

28. Establish a mechanism, such as a system safety status board, to provide visibility to management of the status of the system safety program.

29. Issue a directive and implementing instructions under program management signature establishing membership and operating instructions for an accident/incident investigating committee.

30. Revise the system safety checklist based upon any modifications made to the system safety requirements brought about during this phase of the program.

ENGINEERING

1. Determine what engineering system safety tasks are called for explicitly or implicitly in response to the contract work statement, and ensure that appropriate requirements and tasks are established in the system safety plan.

2. Establish a separate and identifiable organization in engineering, for performing engineering system safety tasks determined to be necessary.

	Percent Complete	Responsible Party

3. Establish a head of the engineering system safety function and define authority and responsibilities for that person.

4. Issue directives necessary to ensure that the head of engineering system safety has access to the chief engineer on major safety problems.

5. Issue engineering implementing instructions to permit system safety functions and responsibilities to be carried out effectively.

6. Arrange for system safety participation in engineering design reviews and change board activities.

7. Identify system safety tasks required in the engineering activities to be carried out during the system life cycle.

8. Arrange for system safety tasks and functions to be included in appropriate engineering program schedules and milestone charts.

9. Review system safety audits and surveys of the system, equipment, and facilities for their interfaces with engineering activities and functions.

10. Arrange for system safety progress reports, status reports, and briefings to be made to the program manager and to the customer.

11. Ensure that system safety problems and recommended corrective actions are reviewed as a regular agenda item at engineering staff meetings.

	Percent Complete	Responsible Party

QUALITY ASSURANCE

1. Ensure that evacuation plans have been prepared for each hazardous area and for each hazardous activity that permit expeditious evacuation of personnel.

2. Ensure that egress provisions such as doors, stairways, and elevators are adequate.

3. Check for the existence and adequacy of emergency plans to be activated in the event of catastrophic events, such as blasts or explosions, leaks or spills, fires, extreme weather, earthquakes, civil disturbance, etc.

4. Check for inclusion in these plans of the procedures to be employed for extinguishing different types of fires that may result in the various manufacturing areas.

5. Ensure that emergency plans and procedures are posted in prominent areas and that they are updated as necessary.

6. Check that personnel receive proper training for carrying out emergency procedures.

7. Determine that the number of drills scheduled to simulate emergency hazardous conditions is adequate.

8. Ensure that responsibility is established for writing and updating emergency plans and procedures for all safety-critical activities.

9. Ensure that emergency crews, such as medical personnel, decompression chamber operators, and fire crews, are on station as required during safety-critical operations.

	Percent Complete	Responsible Party

10. Determine the adequacy of warning systems such as lights, loudspeakers, and sirens, and ensure that the operability of such systems is checked on a routine basis.

11. Ensure that systems used to detect fire, toxic gas, combustible fumes, carbon dioxide, etc., are installed at appropriate locations and are inspected routinely for operability.

12. Ensure that personnel use protective clothing while conducting hazardous operations, in accordance with safety standards established for this purpose.

TESTING—OPERATIONS PHASE

1. Ensure that all facility, medical, and other personnel supporting testing receive a safety-oriented training course and periodic updating.

2. Ensure that all personnel receive indoctrination concerning hazards, safety rules, and emergency instructions for each safety-critical activity they perform. Update such indoctrination as appropriate.

3. Determine that certification is required as appropriate for test personnel and for each group carrying out critical tests. Ensure that testing for renewal of such certification is performed routinely.

4. Establish on-the-job safety training for personnel assigned to test operations prior to assumption of their duties.

	Percent Complete	Responsible Party

5. Ensure that personnel are briefed on potential hazards associated with each safety-critical task.

6. Ensure that key personnel attend demonstrations of hazardous events and appropriate responses to their occurrence are demonstrated.

7. Determine that drills on simulated emergency hazardous conditions are scheduled routinely.

8. Ensure that all test personnel attend regular safety meetings.

9. Ensure that supervisor training programs exist to instruct appropriate reporting of accidents and safety problems and for carrying out corrective action required to remedy safety discrepancies or problems.

10. Ensure that a glossary of standard terms is available for use during safety-critical tests to preclude potential misunderstanding during communications carried out during test activities.

11. Ensure that temporary personnel, visitors, outside contractors, etc., are aware of safety regulations.

12. Conduct unscheduled checks on personnel and equipments to determine the extent of conformance to all safety regulations.

SYSTEM SAFETY PROGRAM OUTLINE

The system safety program plan (SSPP) is the basic document, usually a part of a contractual agreement, for specifying the system safety effort. It describes, for example, the safety level required, how and on what schedule the requirements are to be met, and who shall perform the various tasks needed to achieve the required safety level. This appendix provides an outline for preparing a SSPP.

1. **General Considerations**
   - 1.1 Introduction
   - 1.2 Scope and purpose
   - 1.3 Application and implementation
   - 1.4 Applicable documents
2. **Safety Organization, Responsibilities, and Authority**
   - 2.1 Relationship to total organization
   - 2.2 Organizational array
   - 2.3 Responsibilities and authority
   - 2.4 Interface relationships
   - 2.5 Associate contractor organizational arrays, responsibilities, and authorities
   - 2.6 Subcontractor organizational arrays, responsibilities, and authorities.
   - 2.7 Organizational arrays, responsibilities, and authority of system safety working groups
3. **System Safety Criteria**
   - 3.1 Definitions
   - 3.2 Hazard level categories
   - 3.3 System safety precedents
   - 3.4 Special contractual requirements
   - 3.5 Identification of techniques for analysis

4. **System Safety Tasks to Be Completed**
   4.1   Qualitative analyses
   4.2   Quantitative analyses
   4.3   Operating analyses
   4.4   Program review participation
   4.5   Design review participation
   4.6   Contractor audit activities
5. **System Safety Documentation**
   5.1   Number, format, and schedule of reports to be prepared for submission to others by system safety
   5.2   Number, format, and schedule of reports to be received for use by system safety
   5.3   Forms for reporting potential hazards
   5.4   Procedures for accident investigation
   5.5   Safety bulletins
   5.6   Safety data bank
   5.7   Procedures for disseminating safety data among associates and subcontractors
   5.8   Medical reports
6. **Safety Program Control**
   6.1   Intermediate and final milestones of tasks to be conducted
   6.2   Procedures and schedule for safety audits of subcontractors
   6.3   Procedures and schedule for internal safety audits
7. **Safety Training**
   7.1   Personnel qualifications, including medical, for hazardous activities
   7.2   Training and certification programs for manufacturing, test, maintenance, and quality assurance personnel and for system operators and users
   7.3   Training for response to emergencies
8. **Facilities and Support Functions**
   8.1   Requirements and procedures for handling and storage of potentially hazardous material

GLOSSARY

Throughout the text, terms pertinent to system safety have been defined. In the Glossary the most general and useful ones in qualitative and quantitative analyses have been grouped as a handy reference for the reader. For specific needs, the reader should refer back to the text. Highly technical terms that apply only to the reader's field or to special instances may not be found here, but the reader should feel free to add those definitions.

Accident. An unplanned event or series of events that results in death or major injury to personnel or damage to equipment, experiments, associated support equipment, or facilities.

Accident Risk. Measure of vulnerability to loss, damage or injury caused by a dangerous element or factor. (A probabilistic expression.)

Accident Risk Factor. A dangerous element of a system, event, process, or activity, including causal factors such as design or programming deficiency, component malfunction, human error, or environment which can transform a hazard into an accident.

Audit. A methodical examination and review to verify satisfactory achievement of some aspect or an entire safety effort.

Contractor. A private-sector enterprise or organizational element of government engaged to provide services or products within agreed limits specified by the managing activity.

Credible Condition. A condition that has a reasonable likelihood of occurrence.

Criticality. A quantitative assessment of importance that is a function of the probability of occurrence of a hazard and the severity of the hazard's effects.

Damage. Breakage or ruin of hardware or software by causes transmitted across system or component interfaces by inadvertent internal or external action including component failure and human error. Damage can cause obstruction of critical functions or cause the need for repair or replacement.

Deviation. An alternate method of compliance with the intent of fulfilling specific requirements.

Hazard. An existing or potential condition whose occurrence can result in a mishap. (The presence of fuel in an undesired location is a hazard whereas the fuel itself is not.)

Hazard Probability. The likelihood, expressed in quantitative or qualitative terms, that a hazard will occur.

Hazard Severity. A qualitative assessment of the worst potential consequence, defined by the degree of injury, occupational illness, property damage, equipment damage, or environmental harm, that could ultimately occur.

Line Program Organization. The head of an office, the program office, or the organization responsible for the management of a given operation or task.

Major Injury. Any injury of personnel that results in admission to a hospital, such as bone fracture, second- or third-degree burns, severe lacerations, internal injury; severe radiation or chemical exposure; unconsciousness; severe physical damage; or irreversible injuries of any sort.

Managing Activity. The organizational element that plans, organizes, directs, contracts, and controls tasks and associated functions throughout one or more of the life cycle phases of the system.

Mishap. An unplanned event or series of events that results in injury, illness, or death of personnel, damage to or loss of equipment or property, or environmental harm.

Mission. The specified task or function of a system for each of its life cycle phases.

Program Safety Requirements. The set of contractually imposed design and operational requirements listed in compliance documents or system specifications for a particular project. (Program safety requirements define system constraints and capabilities, establish acceptable and unacceptable risk conditions, or identify specific design and operational criteria and approaches.)

Qualified System Safety Engineer. A technically competent individual who is educated at least to the bachelor of science level in safety engineering, related engineering, or applied science and is registered as a professional engineer in one of the states or territories of the United States or has equivalent experience approved by the managing activity. This individual shall have been assigned as a system safety engineer on a full-time basis for a minimum of four years in at least three of the following six functional areas: (1) system safety management, (2) system safety analysis, (3) system safety design, (4) system safety research, (5) system safety operations, and (6) accident investigation.

Risk. A measure of the combination of the probability and the consequences of the hazards of an operation expressed in qualitative or quantitative terms.

Safety. Freedom from conditions that can cause injury, illness, or death to personnel, damage to or loss of equipment or property, or environmental harm.

Safety Analysis. A document prepared to systematically identify the hazards of a system, to describe and analyze the adequacy of the measures taken to eliminate, control, or mitigate identified hazards, and to analyze and evaluate potential accidents and their associated risks.

Safety Concerns. Those identified safety-critical aspects or risk factors which cannot be satisfactorily resolved or closed out by the contractor and must be brought to the attention of the managing activity for resolution.

Safety-Critical. Used to describe any condition, event, operation, process, equipment, or system with a potential for major injury or damage to people, hardware, software, or the environment.

Subsystem. An element of a system that, in itself, may be considered to be a system for purposes of safety analysis.

Support Equipment. All equipment required to support the system mission during each of its life cycle phases. Examples of support equipment include aerospace ground equipment (AGE), maintenance ground equipment (MGE), transportation and handling (T&H) equipment, and equipment used to support system deployment (e.g., assembly tools and fixtures, test and checkout equipment, personnel support and protection equipment).

System. A composite, at any level of complexity, of personnel, materials, tools, equipment facilities, environment, and software. The elements of this composite entity are used together in the intended operational or support environment to perform a given task, to achieve a specific production, or to support a mission requirement.

System Loss. The occurrence of damage to the system sufficient to render repair impractical so that salvage or system replacement is required.

System Safety. The optimum degree of safety within the constraints of operational effectiveness, time, and cost attained through specific application of system safety management and engineering principles whereby hazards are identified and risk optimized throughout all phases of the system life cycle.

System Safety Engineer. See *Qualified System Safety Engineer.*

System Safety Engineering. An element of system engineering requiring specialized professional knowledge and skills in applying scientific and engineering principles, criteria, and techniques to identify, eliminate, or control system hazards.

System Safety Management. An element of program management that establishes the system safety program requirements and ensures the planning, implementation and accomplishment of tasks and activities to achieve system safety consistent with the overall program requirements.

System Safety Organization. A group of people responsible for establishing, managing, and performing an overall system safety program.

System Safety Program. The combined tasks and activities of system safety management and system safety engineering that enhance operational effectiveness by satisfying the system safety requirements in a timely, cost-effective manner throughout all phases of the system life cycle.

System Safety Program Plan (SSPP). A formal document that fully describes the planned safety tasks required to meet the system safety requirements, including organizational responsibilities, methods of accomplishment, milestones, depth of effort, and integration with other program engineering and management activities and related systems.

Absolute value, 66–67
Absorption law, 54
Air Force Systems Command, 39
Algorithm
  bottom-to-top, 199–200
  computer solution, 199
  cut set, 199–211
  top-to-bottom, 200–211
AND gate, 167
  priority, 173–175, 174
  qualified, 215
  sequential, 174
Arborescence, 159
Arc, 157
Associative law, 52
Automobile safety, 2, 27
Availability, 57–58

Bacon, Sir Francis, 71
Bay Area Rapid Transit, 74
Bernoulli mass function, 274–275, 280
BICS, 201–211
Binary string, 228–230
Binomial density function, 269, 271–272
BIS, 207–208
Bit fiddling, 225–230
Boole, George, 52
Boolean algebra, 50–55
Boolean indicated cut sets, 201–211
Boolean indicated sets, 207–208
Bottom-to-top algorithm, 199–200
Branch, 157
Bundling, 222–225
  event, 222–224
  gate, 225

Capacity, of fault tree, 159
Chain, 157
Chance failure, 265–267
Chebychev, P. L., 282
Chebychev's inequality, 281–288
Checklists, 106–109
Circle (end event), 163

Classification of hazards, 12–13,
  147–149, 252–253
Combinations, 219
Combinatorial AND gate, 174
Combined hazards, 126–127
Common cause failure, 230–233
Commutative law, 52
Complement, 50–51
Component importance, 250
Compound event, 44
Concept formulation phase, 83–84
Conditional probability, 243–244
Connected graph, 158–159
Consumer Product Safety Commission, 2
Continuous density function, 267–271
Contract definition phase, 85–86
Cost of quality, 65
Cost value, 20
Criteria for system safety, 81–82
Critical path, 111
Criticality, 251–253
Cumulative distribution function, 261
Current indicator (of hazards), 123
Cut set, 197–211
  algorithm for, 199–211
  importance of, 230, 250
  minimal, 197–211
Cycle, 158

De Morgan's law, 54
Decision (structure), 94–95
Demoivre, Abraham, 267
Density function, 261–265, 268–269
  continuous, 267–271
  discrete, 271–276
Dependent events, 241, 243
Development phase, 86–87
Diamond (end event), 164–165
Discrete density function, 271–276
Disposal phase, 88
Distributive law, 52
Distribution function, 261
Documentation for safety, 90–91